Managing Concurrent Engineering

Managing Concurrent Engineering
Buying Time to Market

A Definitive Guide to Improved
Competitiveness in Electronics
Design and Manufacturing

Jon Turino
Logical Solutions Technology, Inc.
Campbell, CA

 VAN NOSTRAND REINHOLD
_____ New York

Library of Congress Catalog Card Number 91-34680
ISBN 0-442-01078-8

Manufactured in the United States of America

Van Nostrand Reinhold
115 Fifth Avenue
New York, New York 10003

Chapman and Hall
2-6 Boundary Row
London, SE 1 8HN, England

Thomas Nelson Australia
102 Dodds Street
South Melbourne 3205
Victoria, Australia

Nelson Canada
1120 Birchmount Road
Scarborough, Ontario M1K 5G4, Canada

16 15 14 13 12 11 10 9 8 7 6 5 4 3 2 1

Library of Congress Cataloging-in-Publication Data

Turino, Jon L.
    Managing concurrent engineering: buying time to market /   Jon Turino.
        p.   cm.
    Includes bibliographical references (p.   ) and index.
    ISBN 0-442-01078-8
    1. Electronic industries.   2. Concurrent engineering.   I. Title.
TK7836.T87   1982
    621.381—dc20   91-34680
                                                              CIP

This book is dedicated to my clients and customers—past, present, and future—who keep asking the hard questions.

And to Tina, who put up with its being written and whose review and advice have significantly continuously improved it.

Contents

Preface

This book is one of the first texts to emerge on the topic of concurrent engineering that is aimed specifically at the electronics design and manufacturing marketplace. It is, in essence, a distillation of the knowledge that I have been privileged to gain over 12 years of consulting and seminar leadership literally all over the world. It is also based on the practical experiences of companies that have embraced and, more importantly, are still embracing the continuous design improvement process known as concurrent engineering.

There is no magic in implementing concurrent engineering. What it takes is education, commitment, and cooperation. It is the aim of this book first to educate managers and engineers alike in both the organizational and technical issues involved in successfully implementing concurrent engineering. For only with education will understanding be forthcoming.

It is possible to reduce time to market, improve both design and product quality, improve company profitability and competitiveness, and increase customer satisfaction through the adoption of concurrent engineering. The examples presented in Chapter 2 are but a sampling of the tremendous gains that have been (and continue to be) achieved. It is the second aim of this book, therefore, to evoke a commitment on the part of top managers to a company-wide culture change to the concurrent engineering process so that their organizations can achieve similar competitive advantages.

The third aim of this book is to engender cooperation between the various functional elements present in most electronics design and manufacturing companies. This will be done by demystifying the elements of and techniques for implementing concurrent engineering, and by presenting, in clear and simple terms, the basic principles of design for manufacturability, design for testability, and design for serviceability.

Concurrent engineering is one of the ways to achieve total quality success in both product and process design and operations. In that respect, it is both as difficult—and as rewarding—as total quality management in the manufacturing operations of an organization. Concurrent engineering is, in fact, part of the overall structure required for a total quality management organization.

It is my sincere hope that this book will provide you with the insights and knowledge that you need to succeed in the increasingly globally competitive marketplace of the 1990s. For if it does, I will have succeeded also and will continue to chronicle your successes in our ongoing Concurrent Engineering Seminar™ series.

Thank you, my readers, for acquiring what I hope will become your "bible" on concurrent engineering in electronics.

Jon Turino

Acknowledgments

This book holds the accumulated knowledge of a great many people with whom I have become acquainted and, many times, friends with, over the past 20 years. All deserve thanks and acknowledgment, and some deserve special mention.

Among those deserving special mention are some of the people who attended the first Concurrent Engineering Seminar™ sessions that began in early 1990. They had the desire to learn more about concurrent engineering and the faith that I knew what I was talking about. Their feedback was invaluable in both validating the concurrent engineering message and shaping the format of this book. A short (and unavoidably incomplete) list of those who I wish to acknowledge includes:

Brian Trexel, Associate Director of Engineering, ComStream Corp.
Jim Lantz, Manager of Electronic Design, Delco Systems
Gene Davenport, Manufacturing Engineer, Pacific Scientific
Robin Savage, Manufacturing Engineering Manager, Comstream Corp.
Farzin Kamkari, Project Engineer, Pacific Scientific
David Keezer, Test Engineering Manager, Comstream Corp.
Greg Smith, Senior Test Engineer, Apple Computer Inc.
LeRoy Leland, Manager of Engineering Projects, Itron
Steve Ennis, Manager, Electronic Design Services, Northern Telecom
Lorin Rocks, Maintainability Engineer, Lockheed Missiles and Space Company
David Shawbitz, Functional PCB Test Lead Engineer, Maxtor
Steve Anderson, Principle Test Engineer, Prime Computer
Marshall Hudson, Director of Engineering, Telco Systems
Stephen Wilson, Product Design Supervisor, Anaren Microwave
Richard Barnello, Product Design Engineering Manager, Anaren Microwave
Wayne Keevan, Test Engineering Section Manager, Prime Computer
Nihar Mohapatra, Principal Hardware Engineer, Prime Computer
Tony Hudson, Production Test Engineer, Product Introduction, Prime Computer

Thanks, people. You were the pioneers. Many have followed you, and I thank them as well. Although the names are too numerous to list, each of you know who you are and hopefully know that you also have our thanks. May the sun continue to shine on your face and the wind be at your back.

Managing Concurrent
Engineering

1

Introduction to Concurrent Engineering

Increasing pressures in the marketplace for electronic products are forcing changes in the way organizations develop new products. There is increased pressure to get products of ever higher quality to customers in ever shorter times. Product life cycles are decreasing as well, and product price/performance ratios are being scrutinized more carefully. The traditional "serial" approach to product design and development (see Figure 1-1) used by many companies today is, therefore, hampering their ability to compete effectively in what is becoming an increasingly global marketplace for electronic (and other) products.

In a serial engineering environment, design is often done in a relative vacuum.

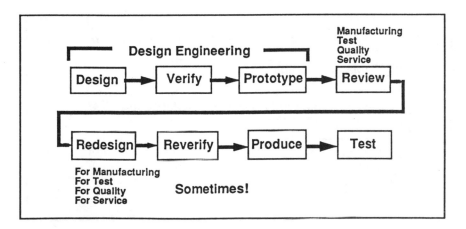

FIGURE 1-1   The serial design process

Manufacturing, test, quality, and service organizations may not see a design until it is virtually completed. If they raise points during design reviews regarding the difficulty, time, and expense involved in producing the design as presented, they may cause the need for the product to be redesigned. This redesign activity adds to overall time to market for the product. If a redesign is too expensive or too time consuming, no action will be taken to improve the manufacturability, testability, quality, or serviceability aspects of the product, and it will be more expensive to produce, verify, and support than it could (or should) be for its entire life. All of these factors hamper competitiveness.

One solution to improving competitiveness is to change from a serial design and development process to a concurrent engineering process (see Figure 1-2). Instead of allowing design groups to develop products that must be "force fitted" into existing manufacturing, test, and service processes, with no forethought to such product attributes as design for manufacturability, design for testability, or design for serviceability, smart managers are converting to a design team concept that moves previously "downstream" product considerations "up front" (i.e., right from the start of the product development effort).

The concurrent engineering process treats design for manufacturability, testability, quality, and serviceability attributes (among many others) equally and in parallel with product design for performance attributes such as speed, power consumption, size, weight, and reliability. It also provides early visibility for any changes to manufacturing, test, or support processes that may be required in order to handle new product design, fabrication, packaging, or test technology features that will make the product itself more competitive.

Concurrent engineering integrates the expertise from all of the various engi-

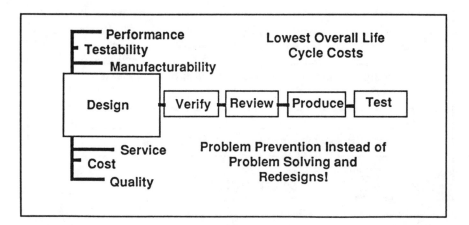

FIGURE 1-2   The concurrent engineering process

neering disciplines during the actual design phase. Tradeoffs regarding producibility, testability, and serviceability are made in real time. With proper forethought, many of the problems that typically plague an organization using the serial engineering process can be completely prevented. Then, when a design is verified, it is already a manufacturable, testable, and serviceable design of high quality, and the design review is held just to make sure that nothing was missed in the process.

The whole focus of concurrent engineering is on a "right-the-first-time" process, rather than on the typical "redo-until-right" process that is so common in the serial engineering mode of operation. The elimination of design iterations reduces product development costs and shortens time to market for new products.

DEFINITIONS AND CONCEPTS

It is helpful when dealing with any concept to provide a formal definition for it. The definition used for concurrent engineering throughout this text is:

> Concurrent engineering is a systematic approach to the integrated, simultaneous design of both products and their related processes, including manufacturing, test, and support.

Concurrent engineering is a concerted corporate effort to achieve maximum efficiency, economy, and quality throughout the total business cycle—from product concept through design, verification, manufacture, test, and service. Thus, we are talking *teamwork, communication, culture, commitment, customer satisfaction, company competitiveness,* and *early attention to manufacturing, test, and support issues.* This should not occur after the fact, where redesign is time consuming and expensive, nor should it be ignored, which means paying for the lack of proper concurrent engineering design over and over again in the factory and in the field.

It is also important to recognize what concurrent engineering is not. It is not an attempt to restrict design innovations nor to criticize the ability of a design (or a designer!) to perform the required functions. No one is trying to stifle the product designer's ability to be creative and innovative in designing new products. The idea is to insure that the best products, from the technology and features standpoints, can be produced, verified, and supported in the shortest amount of time at the least overall business cost. In a concurrent engineering environment, designers can be (and are encouraged to be) as creative as they want when creating products that customers need that are also manufacturable, testable, and serviceable.

Another way to look at the concurrent engineering concept is to contrast it with the traditional design environment (see Figure 1-3). This figure represents the antithesis of the simultaneous design of both product and process that is the embodiment of the concurrent engineering process.

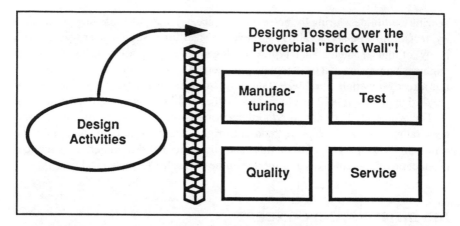

FIGURE 1-3   The traditional design environment

The concurrent engineering process, in contrast, makes use of "product birthing teams." Such a team, sometimes also called a new product development team, or a time-to-market team, is illustrated in Figure 1-4.

The successful concurrent engineering team is multifunctional and multidisciplinary. Each member contributes his or her expertise to the overall product design, and each is responsible to the other team members for the timely and cost-effective delivery of the product to the customer. The figure is meant to be illustrative, of course, in that it does not list all of the possible disciplines that may be required for a specific project. A fairly complete list of potential team members is contained in the Concurrent Engineering Team Checklist (see Appendix). So the figure below represents the minimum team makeup, not the maximum.

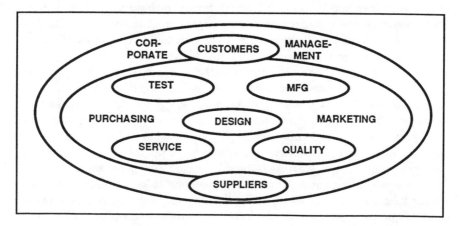

FIGURE 1-4   The product birthing team

Notice specifically that customers and suppliers are included in the makeup of the team. Close customer contact during product design is essential if the product eventually designed is to fit the customers' needs and therefore be successfully marketed. Suppliers are also included, since they can often offer new components or technologies that many times help to lower product costs, improve product performance, or reduce design time. Management's job in the concurrent engineering environment is to provide the *right resources with the right information at the right time*, to facilitate optimum team performance.

It is also useful to define the terms design for manufacturability, design for testability, and design for service. The definitions used in this text are as follows:

- Design for manufacturability is a measure of the ease with which a product can be assembled in the shortest amount of time and with the least amount of assembly labor, without introducing defects into it.
- Design for testability is a measure of the ease with which a comprehensive test program can be created and executed, as well as the ease with which the item can be tested and diagnosed in the shortest amount of time while using the lowest cost testing resources.
- Design for serviceability is a measure of the ease with which the status (operational, nonoperational, or degraded) of a product can be determined, as well as the ease and speed with which it can be returned to full operational status in the event of a fault.

These definitions are specifically broad so that they may be applied across the board to virtually any organization, or product, in that organization's own language and acronyms. They do, however, encompass all of the principles whose implementation details are covered in Chapters 7, 8, and 9.

ELEMENTS OF CONCURRENT ENGINEERING

Concurrent engineering, as used in this text, is made up of a minimum of five specific design activities that go on in parallel. These five activities are:

- Design for performance (DFP)
- Design for manufacturability (DFM)
- Design for testability (DFT)
- Design for serviceability (DFS)
- Design for compliance (DFC)

Of course, many other specific factors may enter into the product design equation, including such things as assembler and operator safety, industrial or packaging design, environmental hazard considerations, thermal analysis, and mechanical design (to name just a few). These additional factors may either be considered as subsets of the five main design activities associated with the concurrent design process or as additions to them. The point is to identify, up front, all of the expertise required to meet all of the product requirements and to make sure that nothing "falls in a crack."

Design for performance issues include making sure that all customer-required functions are implemented, that the product operates at the required speed, that human factors such as ease of operation and safety have been taken into account, and that reliability goals have been met. A product that does not perform a useful function is a useless product that will not sell. Close relationships with customers are critical in identifying the functions that are really required in a product. And adding "bells and whistles" just because they can be added can actually hurt a product. The customer may perceive it not as "more powerful" but as "more complex"! Furthermore, adding additional *useful* functions can give a product a marketing edge. Adding functions just for the sake of adding them, however, can make the product more difficult to produce, verify, deliver, operate, and service.

Design for manufacturability issues include using the minimum number of parts in the product (consistent with other tradeoffs), facilitating error free assembly, using standard components whenever possible, and fitting the product design into the process that will be used to produce it. Even where flexible manufacturing systems exist, minimizing the number of design variations will reduce setup times, cycle times, and inventory levels.

The basic design for manufacturability principles can be basically described as "keep it simple, s_____"! There is still an attitude of "Our job is design—let someone else worry about building it!" that is far too prevalent in most organizations today. Facilitating error free assembly is a case in point. When one company involved with a Japanese joint venture achieved a first pass yield of 90 percent on an initial production run of disk drives, the U.S. engineers went off to celebrate. The Japanese scratched their heads glumly, trying to figure out why the yield was so low. Concurrent engineering means being aware of manufacturing constraints and making allowances for them so that very high yields can be achieved.

Design for testability has been talked about for over 30 years and is still perceived as being difficult to implement, detrimental to product performance, and too expensive in terms of parts cost and product "real estate." Those perceived barriers to good testability may have been partially true years ago, but in this day and age of high complexity, high technology devices, boards, and systems, they are categorically untrue.

The three basic testability axioms—partitioning, control, and visibility—are easy to implement, and there are many new techniques for implementing them available, including automatic synthesis in the CAE environment. Gates are extremely inexpensive at the silicon level, and pins are getting less expensive as device pin counts increase.

Facilitating automated testing and including a built-in test may not be trivial, but they are no longer arduous either. The best design, from a "functions per square something" standpoint, is absolutely useless unless its functionality can be verified (i.e., it can be tested) in a timely manner at a reasonable cost.

Design for quality includes the basics of meeting customer quality expectations. It include things like selecting high quality components in the first place (regardless of minor pricing variations), using proper derating calculations to insure reliability, and making sure that the product has an attractive "look and feel." They've been well documented, and the techniques for their implementation are the subject of many seminars and programs, including those by Crosby, Deming, and others.

Concurrent engineering is actually just an extension of design for quality where one looks not only at the "external customer," or end user of the product being designed, but also at the "internal customer"—the people and organizations to which designers provide their design "product." High quality is a design requirement. Products without quality designed in will eventually fail in the marketplace. They may fail to even reach the marketplace. And what good is a design that never sees the light of successful implementation and use?

Things break. According to Murphy's Law, they always break at the most inopportune time and in the most inopportune way. And someone has to fix them. Concurrent design efforts go a long way toward determining the level of ease or difficulty with which we can fix a nonoperating product.

Design for serviceability principles include things like making replaceable items accessible, replacing fuses with circuit breakers, partitioning designs into modular functions (both electrical and mechanical), and including built-in test and remote diagnostics as appropriate to the application. Are they really all that tough to implement? Will they really cause "the schedule" to slip? Will they really "cost too much"? If so, what is the cost—parts cost, design cost, assembly cost, service cost, or *overall business cost*? Many companies are saving millions of dollars per year in service costs by increasing hardware costs 10 percent or more. They're doing so by making the right serviceability tradeoffs *during design*!

Design for compliance with the various regulatory agencies around the world is another area that must be considered. Thus, team members with expertise in things like the FCC (Federal Communications Commission), UL (Underwriters Laboratory), CSA (Canadian Standards Authority), and others, including special medical instrument requirements, special requirements in Japan, and the emerging standards being developed as 1992 approaches in Europe, may also be required as part of the concurrent engineering effort.

It is sad, indeed, to find out that a product won't meet regulatory requirements after the design is complete. This may entail added costs, for example, for extra shielding (for RF emissions) or a complete change in packaging materials or methods. That kind of a retrofit design iteration can really impact time to market.

There are no hard and fast rules for implementing all of the concurrent engineering guidelines contained in this text except one: Thou shalt design concurrently!

Few design engineers were taught the nuances of manufacturing, test, and service during their university education. Also, few designers design difficult to manufacture and difficult to test products on purpose—they do it out of lack of knowledge of the processes that their designs must undergo in order to see the light of the marketplace. The keys are to be conscious of that lack of knowledge, employ the knowledge of others to supplement specific knowledge weaknesses, and get everyone as trained and educated as possible. In management, this is known as "hiring to strengthen your weaknesses."

Not everything contained in the technical chapters of this book can always be implemented. But it can always be considered, and the proper tradeoffs can be made during design.

IMPACT OF CONCURRENT ENGINEERING

Concurrent engineering is a culture change for many organizations. It thus has a big impact on each of the people in the organization and on how they operate. The impact of concurrent engineering on designers is significant. Where once design could be done "in a vacuum," without regard to the processes by which the design would be manufactured, tested, and supported, with concurrent engineering all of these previously back-end activities move to the front of the design cycle.

Concurrent Engineering is intended to cause designers, from the very beginning of a design activity, to consider all elements of the product life cycle, from product concept through design, manufacture, service, and even disposal, including quality, overall business costs, time to market, and customer needs.

For management, the change to a concurrent engineering process and culture requires new leadership skills and methods. Managers must become adept at breaking down barriers in the organization and leading and coaching team leaders and team members as a design begins, evolves, and progresses. They must provide the teams with the data and tools that they need to design quantitatively, and they must wholeheartedly support the concurrent engineering process. They must also be prepared to intervene if the process breaks down within a team or if one function of the team is dominating the tradeoff interactions, reminding them that the ultimate decision criteria is customer requirements.

Concurrent engineering is intended to foster team leadership, in addition to functional management, in order to provide the right resources and expertise at the right time and in the right place, on a project and process-oriented basis.

WHY DO CONCURRENT ENGINEERING?

Concurrent engineering is important because the design activity has a very significant impact on the cost of producing, testing, and servicing a product. Decisions made during the design phase of a product have the biggest impact on overall product cost over the life of a product. To compete effectively in world markets in the 1990s, costs for all elements of electronic design, manufacturing, test, and service must come down.

Figure 1-5 shows the typical relationship between the cost to detect, isolate, and repair a defect at the component, printed circuit board assembly, final product (or system)level, and in the field.

An interesting phenomenon occurs with respect to the cost of detecting, diagnosing, and repairing a defect at succeeding levels of product integration. As shown in the chart below, the cost increases by a factor of ten (i.e., an order of magnitude) at each succeeding level.

As part of the manufacturing test process, a critical objective is to make sure that defective components do not end up on boards, where they are much more expensive to find and replace. Similarly, very high board level fault coverage is needed so that defective boards do not go into systems. Finally, of course, we do not want defective products to leave our factories. If they do, customers get to find them. This is not only expensive in terms of actual service cost, but also in terms of company image, customer satisfaction, and, ultimately, market share.

Thus the objectives are to (a) prevent defects from occurring in the first place and (b) test for defects as early in the manufacturing process as possible.

A similar phenomenon occurs in the development of a new product. Making design changes gets much more expensive as the design progresses, with the earliest decisions having the largest impact (see Figure 1-6). Not only does the same basic relationship hold, but the figures themselves are orders of magnitude higher!

That's why it is important to consider all of the aspects of both the product and the process when the concept (or specification or requirements) for the product are

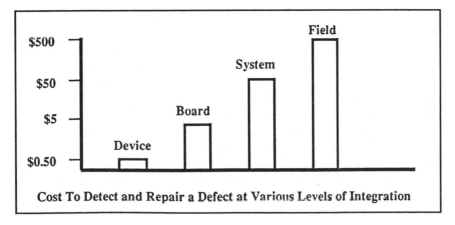

FIGURE 1-5   Manufacturing, test, and service economics

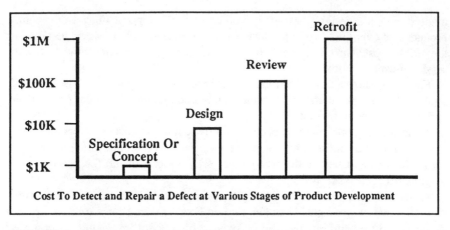

FIGURE 1-6   Concept, design, review, and retrofit economics

first set out and why designers need to be educated so that they are aware of the concurrent engineering technical guidelines and trade-offs while they are doing the actual design job. Then, design reviews are used to make sure that nothing was missed and to prevent the need for design iterations (i.e., a "retrofit").

The objectives of concurrent engineering are to achieve:

- Shorter time to market
- Lower product development costs
- Higher product quality
- Lower manufacturing costs
- Lower testing costs
- Reduced service costs
- Enhanced competitiveness
- Improved profit margins

There is great pressure to develop new products in the shortest amount of time in order to meet the ever-changing demands of the marketplace. Product development times are not the only item being affected by changing technology. Product life cycles are becoming shorter as well. This means that flexible manufacturing methods and quicker test methods must be developed and implemented. Product designers have a big impact on the effectiveness and timeliness of the implementation of modern production methods.

Shorter time to market is the most widely used phrase to describe a competitive enterprise today. Beating the competition, which is becoming ever more aggressive, is becoming ever more important. It means being better at responding to customer needs more quickly. Advances in new technology, if implemented in a concurrent engineering environment, can help us achieve those goals and survive in the 1990s.

Another element in the overall business cycle that needs to be addressed is actual product development costs themselves. Computer aided engineering (CAE) systems have given us the tools to design very complex products very quickly. But proving that those designs will actually work—before committing them to hardware—is becoming both more important and more difficult. Many of the concurrent engineering guidelines presented in this book are aimed at helping not just the manufacturing, test, and service people, but also the design engineer in terms of faster and easier design verification and the minimization of design iterations.

An element that is absolutely critical for survival in the 1990s is improved product quality. Customers are demanding it. Higher product quality, therefore, is just as important as shorter time to market. Higher quality designs lead to higher quality products. And higher quality products can be manufactured with fewer defects, thus lowering the cost of manufacturing and test. The lowest product cost occurs when products are designed and built "right the first time."

Higher quality products also facilitate the use of statistical process control (SPC) and total quality management (TQM) programs, which in turn lead to improved competitiveness due to an improved company image. Product designs must not only meet customer performance and quality requirements, they must also be producible in a competitive manner. That means designing new products with fewer parts (and fewer types of parts) in order to reduce assembly labor costs and achieve production economies.

The fewer the number of defects manufactured into a product, the lower the cost to produce it. Many factories have "factories within factories"—the second factory is in place to repair the defects produced by the main factory! Proper design for manufacturability as part of the concurrent engineering process can reduce manufacturing costs and make production possible in less time.

Testing is becoming an ever larger part of the product cost equation as products continue to increase in complexity. Test program generation times in particular increase rapidly with product complexity. With a little attention to design for testability techniques, weeks (or even months) can be shaved from production test program generation times, thus further improving time to market response for new product designs. With built-in tests incorporated in new product designs, it is also possible to reduce the large expenditures required for powerful Automatic Test Equipment (ATE) systems that can "overcome" testability problems.

While some products are becoming small enough and inexpensive enough to be classified as "throw away" items when they fail in the field, many others are large enough to require service after installation in the customer's facility. Service costs are often a large (and sometimes of unknown quantity) overhead item in many organizations. Part of the concurrent engineering discipline includes attention during product design to the serviceability aspects of a product in order to reduce these large overhead costs.

Competition is increasing on a global scale. The pace of that increase will continue to increase in the 1990s (and beyond). Trouble-free product introductions are therefore even more critically important to creating a good global reputation and to gaining share in new markets. Major competition for U.S. manufacturers will increase from both Europe and the Far East. For European manufacturers, increased competition from the Far East, the United States, and the member countries of a "unified Europe," beginning in 1992, will also be a fact of life. Efforts are under way by U.S. and European manufacturers to penetrate the Far East markets. So no matter where your organization is located in the world, improved competitiveness is a necessity.

One of the most subtle effects, and one of the most damaging in the long term, is that problem product introductions drain resources—both people and money—away from continuous improvement efforts that can make the factory more competitive. Many industry studies have documented the increased profits that can be reaped when the concurrent engineering discipline (or the many subset variants of it) is adopted. It is possible to literally double the profitability of a company by making sure that product designs can be manufactured, tested, and serviced in the shortest amount of time with the least cost. Those improved profits can then be plowed back into product development or used to improve design automation and automated assembly tools.

Can you afford to spend a little extra time and effort during the design phase of a product, if it can double the profitability of your operation? The answer is: You can't afford not to!

SUMMARY

Concurrent engineering is, in essence, the simultaneous design of both products and the processes by which they will be manufactured, tested, and supported. It brings early involvement of many of the organizational functions and "-ility" engineering expertise to bear right from the start of a product design effort, rather than after many critical design decisions have already been made and where design changes can be time consuming and expensive. Concurrent engineering requires that quantitative trade-offs be made during the design phase of a product, with respect to design for performance, design for manufacturability, design for testability, design for serviceability, and design for compliance. The objective is to make trade-off decisions that will benefit both the customer and the producing organization on terms of shorter time to market, higher product quality, lower overall product cost, and increased customer satisfaction.

The later in the design or manufacturing process that a defect is discovered, or the need for a change is uncovered, the more expensive that change or diagnosis and repair will be. Thus, concurrent engineering places great emphasis on making sure that all of the information and expertise needed for a "right the first time" design are in place on the design team and are functioning under good leadership. In addition to customer requirements and product performance features, the design team also focuses on ways to facilitate error-free manufacturing of new products with minimum test and inspection times and costs.

When the concurrent engineering process is in place, small incremental investments in the nonrecurring design effort and cost are made in order to achieve large savings in recurring costs over the life cycle of the product. Concurrent engineering provides a vehicle for increasing market share in an increasingly competitive global marketplace, while simultaneously increasing profit margins for new product designs.

REVIEW

1. When design for manufacturability, testability, and serviceability are secondary considerations, the design approach being used is most likely (a) *a serial engineering approach* or (b) *a concurrent engineering approach.*
2. In a concurrent engineering environment, design for manufacturability, testability, and serviceability are (a) *up-front activities* or (b) *downstream activities.*
3. In a concurrent engineering environment, design tradeoffs are made on (a) *a qualitative basis (i.e., opinion based)* or (b) *on a quantitative basis (i.e., fact or estimate based).*
4. Concurrent engineering (a) *is* or (b) *is not* intended to restrict design engineering innovations.
5. Concurrent engineering is (a) *a design process change*, (b) *an organizational culture change*, or (c) *both.*

Answers: 1: a, 2: a, 3: b, 4: b, 5: c

NOTES:

2

Improving Competitiveness with Concurrent Engineering

Improving competitiveness through the use of the concurrent engineering process for new product development requires that everyone in the organization understand the basics of the cycle through which a product travels, the frequency of activity in each phase of the product cycle, and where the costs associated with each phase are actually determined. In this chapter, we'll look at the product cycle, understanding product costs, real time to market, the data needed to make good quantitative concurrent engineering decisions, and how to achieve maximum business leverage.

THE PRODUCT CYCLE

The overall product cycle in a business is illustrated in Figure 2-1. The design activity is a nonrecurring cost (or at least it is supposed to be!). Products get designed once per product type. They get built once per product as we duplicate the design in manufacturing. They get tested at many levels and must often be serviced in the field. Our objective in concurrent engineering is to make the right decisions during the nonrecurring activity (i.e., design), to maximize our advantages during the recurring activities that may last for years and years.

We refer to this as creating maximum leverage. Not in terms of borrowing huge sums of money to leverage a buyout of an organization and use its future profits to pay back the debt, but in terms of investing a little time and money during product design to reap larger profits over the life of the product.

As mentioned earlier, maximum leverage cannot be achieved by "retrofitting" a design once it has been discovered that it is difficult, time consuming, and expensive to produce. Some leverage, but not maximum leverage, can be achieved by improving the design using the design review process, but this may still require a design

14

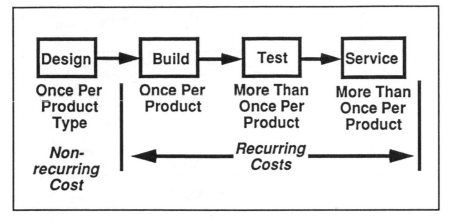

FIGURE 2-1   The product cycle

iteration (either on paper or in software). Maximum leverage occurs with concurrent engineering.

Financial leverage is not the only leveragable gain from concurrent engineering. There is also intense pressure on the design engineering function to bring products to market as quickly as possible. One of the most frequent complaints heard from design engineers is that management imposes unrealistic design schedules. And one of the most frequent excuses for not including concurrent engineering guidelines is that "there is no time—we have to get the product designed as quickly as possible."

This kind of narrow and short-term attitude needs significant adjustment because *time to market is* **not** *just design time!* Time to market is the time it takes to get a product into your customer's hands at a competitive price. If that means having to redesign to lower manufacturing and test costs or to fix performance "glitches" because of inadequate design verification, what is the advantage to rushing a design through? The answer is: There is none.

As illustrated in Figure 2-2, the concurrent engineering discipline is designed to help speed *real* time to market, even if that means spending a little more time making sure the design is flawless in terms of both performance and manufacturability, testability, and serviceability.

Many studies have pointed out the fact illustrated in Figure 2-3—the fact that somewhere between 60 percent to 95 percent of overall product cost is determined during the design phase of a product's life. Product parts costs, assembly costs, and test and service costs are dictated far more often by the design of a product than by the actual manufacturing, test, or service processes and operations. And the earlier the decision, the larger its impact!

The purpose of the concurrent engineering process is to make sure that *the early design decisions can minimize the overall product costs over the life of the product.*

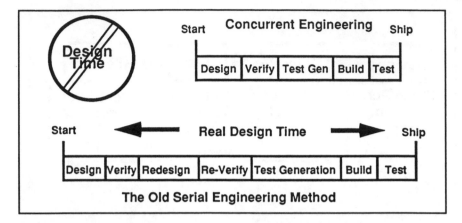

FIGURE 2-2   Real time to market

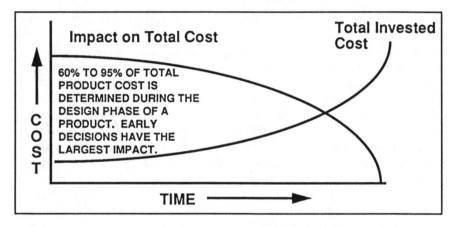

FIGURE 2-3   Impact of design decisions on total product life cycle cost

The early decision to make the product fit into the manufacturing process, rather than requiring manufacturing to acquire new capital equipment, can have a big impact. Taking some extra design time to assure error-free assembly with a minimum number of assembly operations can significantly lower overall product costs and make a company more competitive.

The five most important design for performance issues faced by most designers include product size considerations, weight considerations, speed of operation considerations, human factors, and product reliability goals. Some of the concurrent engineering design guidelines require trade-offs in these factors. It is therefore important to remember that each of these "primary" design factors is but one element in the overall success of a product. When considering the implementation of the concurrent engineering design guidelines, consider them as an additional set of important "design for" items on the list.

Sometimes there are conflicts among just the factors listed above that require early design trade-off decisions. Those decisions have, in the past, often been made without considering the "-ilities." This situation must change if your organization is to be as competitive as it can be in the future in an increasingly globally competitive marketplace environment.

CONCURRENT ENGINEERING BENEFIT EXAMPLES

There are many well documented examples that prove how powerful the concurrent engineering process and culture can be in reducing time to market, improving design and product quality, reducing the number of design iterations, speeding the manufacturing process, and lowering product costs. There is absolutely no reason that your organization cannot achieve similar results through the proper application of the principles explained in this text.

AT&T made several organizational and process changes in the design of the "circuit pack" for its 3B series computer. The "model shop," where prototypes were fabricated for laboratory testing, was completely eliminated. AT&T built the initial units using the equipment that would be used in actual high volume production.

This process, coupled with the extensive use of computer aided design verification and circuit simulation, produced the following results:

- Yields improved from 50 percent to 90 percent
- PCB art master iterations reduced by 33 percent

Achieving these results required actually doubling the size of the up-front circuit design staff. The resulting shortened time to complete each new design, however, coupled with the savings due to improved yields and fewer design iterations, easily paid for the increased up-front costs.

Boeing's Ballistic Systems Division, by using Product Development Teams (PDTs), plus a set of 84 internal improvement initiatives, also achieved some excellent results. Among them were:

- 16 percent to 46 percent cost reductions in manufacturing
- ECOs reduced from 15–20 to 1–2 per drawing
- Inspection costs cut by a factor of 3
- Design analyses for the "-ilities" cut from two weeks to less than an hour
- Material shortages reduced from 12 percent to 1 percent

Each Boeing PDT typically has members from engineering, logistics (test and support), materiel (purchasing), manufacturing, and quality assurance. Each PDT "owns" its product, and each PDT representative has the authority to commit his or her functional organization. Each member of the PDT also participates in and authorizes the release of drawings, request for purchases, and other design and implementation documents.

Burr-Brown chose a concurrent engineering team approach, with excellent results in the design of digital-to-analog and analog-to-digital converters for digital signal processor applications. Personal interaction between the team members yielded better and more manufacturable designs. The process was started by encouraging design, test, and manufacturing inputs during the final revisions of product proposals from marketing and not during final revisions of the product designs themselves. Input continued during design, test development, characterization, prototype production, and device qualification.

Each team member was not only encouraged, but also expected, to ask questions, make suggestions, and offer alternatives throughout the process. The primary team consisted of members from design, test, manufacturing, and marketing, led by a product manager, all dedicated to the project.

Other personnel with additional expertise—purchasing, production, and so on—were called upon as needed during the product design. Weekly meetings were held to reallocate resources as needed on a real time basis so that the next milestone could be met. The result: *time to market cut by six to nine months.*

Codex used electronic integration of the design process to facilitate the concurrent engineering process. Each engineering discipline (e.g., design, test, manufacturing) has simultaneous access to the design representation data base and can thus participate in the design process in real time via his or her workstation.

This investment in networking and computer-aided engineering has *already cut time to market more than 33 percent* and is expected to eventually shave it by a full 50 percent.

Hewlett-Packard (H-P) actually used the Total Quality Control (TQC) process to improve not only its manufacturing performance, but also its administrative and engineering performance. The elements of H-P's program include management commitment, customer focus, statistical process control, systematic problem solving process, and total participation.

Top management commitment in the form of learning, understanding, and leading the TQC efforts with a communicated, unwavering purpose, including ongoing management involvement, is critical. Concurrent engineering uses the same philosophy as TQC—continuous improvement not just of the production process but also of the design process itself. The results for H-P were:

- Scrap and rework costs cut 80-95 percent
- Manufacturing costs reduced by up to 42 percent

- Parts inventories cut by up to 70 percent
- Manufacturing cycle times reduced up to 95 percent
- Product development times cut by up to 35 percent

IBM's East Fishkill facility produces common elements, called masterslices, and other components that become the building blocks for products designed throughout IBM. IBM's concurrent engineering approach centered around the development of an automated electronic design automation system that allowed for simultaneous access to design files by all of the people involved in the design, production and test, and quality processes. Their results:

- Product direct labor costs reduced 50 percent
- Process time for customizing products cut 65 percent
- Three design passes reduced to two (with one as a goal)
- Overall digital electronic design cycle cut by 40 percent

You don't have to be IBM, however, or make huge investments in design automation tools to achieve the same benefits. Human to human communication, while not as fancy, can be just as effective. IBM achieved similarly spectacular results with its famous ProPrinter example. The design was refined to the point where total assembly time was cut from an average of 30 minutes for a previous design to a total of 3 minutes—*a ten to one improvement.*

NCR's 2760 electronic cash register was described in Business Week magazine's May 8, 1989 issue. As the subhead of the article says: "From assembly to accounting, a simple design draws cheers." This particular product has so few "parts" that an engineer can assemble it blindfolded (and did so in front of the press!).

Not all concurrent engineering efforts are this spectacular, of course, but they do serve as models of what can be accomplished. A constant process of achieving small improvements can have a big long-term impact on a company's competitiveness and profitability as well.

The highlights of NCR's concurrent engineering effort on this particular product include the following:

- An 80-percent reduction in parts and assembly time
- 65 percent fewer suppliers
- 100 percent fewer screws or fasteners
- 100 percent fewer assembly tools
- A 44-percent improvement in manufacturing costs
- A trouble-free product introduction

Tektronix made strides similar to NCR's, but in the area of test, rather than assembly. Implementing a complete built-in test and self-calibration scheme as part of the concurrent engineering design process can also pay dividends, as Tektronix proved with a new line of digital storage scopes. Documented in an article in *Electronic Products* magazine, Tek engineers say they cut test-related times by a factor of over 10 (the actual factor is 11.33333—from 8 hours and 30 minutes to 45 minutes total). Those results are remarkably similar to IBM's cut by a factor of 10 in assembly time.

Just imagine what could happen if we could cut both manufacturing and test times by factors of 10.

A final example, although there will have been many more examples documented by the time you read this, should serve to drive home the point that there is a tremendous opportunity to improve design and product quality, product manufacturing processes, and therefore profitability and competitiveness. Table 2-1 illustrates the results that *Texas Instruments (TI)* had with the redesign of a complex infrared sight.

By redesigning the infrared sight (and without reinventing the factory that produced it), TI achieved some impressive reductions in the number of parts, the number of assembly steps, and, therefore, the overall assembly time, which was reduced by a total of 85 percent. Sometimes it even pays to redesign (or retrofit) a product for manufacturability, testability, and serviceability, using all of the resources of a concurrent engineering team.

DESIGN DECISION BASIS

Our experience has been that many product design decisions in organizations that practice serial engineering are made based on opinions, not facts, and that many of those decisions are also made "in a vacuum."

We presented the facts of concurrent engineering via the previous examples. All of the examples are from real companies and have been well documented in the

TABLE 2-1.   **TI infrared sight example**

	Before	After	Reduction %
Assembly Time	129	20	85
Number Parts	47	12	75
Number Steps	56	13	71

trade and business press. *Fact:* Concurrent engineering can get you to market sooner. *Fact:* Concurrent engineering can make you more cost and price competitive. *Fact:* With the complexity of today's product designs and the fickleness of today's markets, you cannot afford to base performance, manufacturability, testability, and serviceability decisions on opinions.

Complexity for complexity's sake is counterproductive. How many of the features of most of your sophisticated electronic products do you (or your customers) *actually use* on a regular basis? Sometimes product simplification can make a product more marketable! That's why you need accurate input from all of the business elements when making design decisions.

Getting closer to customers—with one-on-one meetings between potential product users and the actual product design team—is one way of gathering the facts regarding which design features and parameters are most important to customers. Partnering with customers and suppliers can also help the product birthing team come up with the kinds of quantitative information that they need to make truly informed design decisions.

TIME AND COST DATA USAGE

People in the manufacturing, test, quality, and service functions often have large amounts of data (i.e., facts) regarding the overall times and costs associated with bringing certain products to market (and their ongoing production and warranty/service costs). These facts should be taken into account when new products are designed. We can learn from what we've done right (or wrong!) before.

A word to the wise to those in manufacturing, test, quality, and service functions: The facts you bring forward must be (a) timely, (b) accurate, and (c) presented in the proper manner. The data you hold gives you power. Use it wisely—for improvement, not punishment of other organizations (or, worse yet, specific people in other groups). This means saying "If we do this the same way again, the following will result. If we do it with the concurrent engineering discipline in mind, we have the chance to..." rather than saying "Your last design was a disaster for (manufacturing, test, or service). Look at these numbers..."

The types and granularity of time and cost data required for good concurrent engineering design decisions are illustrated in Table 2-2. For it is still true that if you cannot measure something, you cannot improve it.

Table 2-2 shows detailed breakdowns of each of the major cost elements associated with each major business activity. The design and design verification cost elements are the nonrecurring cost elements in the product development, manufacturing, and service cycle. Depending upon the exact nature of your organization and your products, the list may need to be expanded. Note that these costs need to be estimated for each type of device, board, subassembly, or complete product.

TABLE 2-2.   Summary of cost data needed for concurrent engineering

Design Cost Data for Devices, Bare Boards, Loaded Boards, and Systems	Parts Cost Data
• Design Cost $_____ • Design Verification Cost $_____ • Fault Simulation Cost $_____ • Test Generation Cost $_____ • Iteration Cost $_____	• Glue Logic $_____ • Discretes $_____ • ASICs $_____ • Bare Boards $_____ • Cages/Backplanes $_____
Assembly Cost Data	**Test Cost Data - Capital, Programs, Fixtures PLUS:**
• Board Assembly $_____ • Subsystem Assembly $_____ • System Assembly $_____	• Device Test $_____ • Board Test $_____ • Board Diagnosis $_____ • System Test $_____ • System Diagnose $_____
Quality Cost Data at Each Level	**Service Cost Data**
• Inspection Cost $_____ • Rework Cost $_____ • Escape Cost $_____ • Scrap Cost $_____	• Field Service Call $_____ • Depot Repair $_____ • No Fault Found $_____ • Spares Inventory $_____

Usually, a very small increase in design cost will result in a moderate decrease in design verification cost and a large decrease in fault simulation and test generation costs. Concurrent design also provides the opportunity to actually eliminate the iteration cost. The table also details the cost elements that make up the actual material cost of a product. Here again, depending upon product configuration, the list may need to be expanded, and, as before, the data should be developed for the entire product.

There are occasions where ASICs can be used to replace "glue logic" (and, conversely, where the development of an ASIC is simply not justified). There may be occasions where an increase in the cost of a part (for improved testability characteristics) will be fully offset by a decrease in board, subsystem or system test, and troubleshooting costs. Sometimes, breaking a large board into two smaller (and simpler) boards can make sense. The extra connector cost can be offset by decreased cost for the individual bare boards (reducing the number of layers required, for example).

Assembly cost is another significant element in the cost of a product, depending again upon its size and complexity and the methods used to manufacture it. The costs estimated should include not only the recurring costs at each level of integration but also the nonrecurring cost for capital equipment, machine programming, and the like (amortized over the total estimated number of products of each type to be built).

The recurring test and diagnosis costs for each element of the overall product also need to be ascertained or estimated. Then the "deltas" can be estimated to determine whether design changes for testability are warranted and, if so, just how much testability is affordable, based on potential increased costs for components.

Design improvements may not make a large difference in go/no-go testing costs, but they can make a big difference in troubleshooting (i.e., diagnose) times and costs. The test cost list in Table 2-2 is for recurring test costs. The cost for capital equipment, test programs, and test fixtures should also be estimated and amortized over the total number of items to be built to come up with a "per item" cost that can be used during design to make trade-offs.

Quality costs are another significant element in the overall product cost equation. It might actually be more appropriate to term the costs identified in Table 2-2 as the cost of "not quality," since products that can be produced perfectly every time do not require inspection, rework, or scrap costs! Escape cost refers to the premium paid when a defect "escapes" a test (say at board level) and must be detected, diagnosed, and repaired at a later stage (say at system test) at a much higher cost.

It is also necessary to have yield (or failure rate) and fault distribution figures for each testing and/or inspection step in order to calculate quality costs. Gathering this data, however, can also help in identifying areas where the manufacturing operation itself (without affecting product designs) can be improved to reduce costs and raise quality levels.

Finally, there is service cost data. The list of service cost data identifies the major categories of costs that should be estimated over the service life of a product with the predicted failure rate factored in to come up with a "per item" service cost.

These estimates, along with all of the other elements shown in Table 2-2, should all be taken into account during the design phase of the concurrent engineering process. Only when all of the factors are considered, and all of the proper engineering expertise (i.e., system design, hardware design, software design, PCB design, manufacturing engineering, test engineering, quality engineering, and service, among others) applied, is it possible to develop the best product at the lowest cost in the shortest time.

MANUFACTURING, TEST, AND SERVICE PLANNING

We mentioned the impact of design decisions on product life cycle costs earlier in this text, and the fact that the earliest decisions usually have the largest impact. As more things get cast in silicon (or in copper on epoxy-glass boards or in the final product package), the less our flexibility in changing things to improve the items that make up the elements of concurrent engineering.

Product goals, strategies, and tactics must be planned out as early in the product development cycle as possible—preferably right at the beginning when the product is specified. Those of you in functions that are currently "downstream" from design engineering must take it upon yourselves to get involved in the product design process if you are going to be a source of solutions (rather than an after-the-fact source of problems and complaints).

If the customer requirements dictate a design approach outside the scope of current company capabilities, everyone needs to know about it so that plans can be developed to cope with it. If the design approach can be modified to fit into current company capabilities, so much the better. If neither of those happen, time will be lost, money will be wasted, and market share will be eroded.

Shorter product life cycles and increased pressure for shorter time to market make it imperative that we renounce the "redo-until-right" philosophy and replace it with the "right-the-first-time" philosophy. What does "right" mean? The proper set of design trade-offs for the overall success of the product (and the business), given the specific customer requirements, business capabilities, and competitive environment.

A silicon iteration for an ASIC may cost weeks (or even months) in time to market. An iteration can be caused by a lack of communication between the ASIC designer and the system designer. It can be caused by neglecting to fully simulate the operation of the part in the overall product. It can be caused by having to redesign the part to include boundary scan so that manufacturing can test the product that contains the part. It can be caused by inadequate input from product marketing. There are lots of causes (and even more excuses).

There is only one prevention—concurrent engineering. And even though its practice won't prevent all of the problems all of the time, you have a much better chance to improve your "hit ratio" when you use it properly.

SUMMARY

Time to market can be reduced by as much as 30 percent when the concurrent engineering discipline and process are employed. The chart in Figure 2-4 shows a case where a design iteration was avoided. A close examination of the chart, however, will reveal savings in design verification, test generation, and test, regardless of the design iteration issue. The savings in time to market, even without

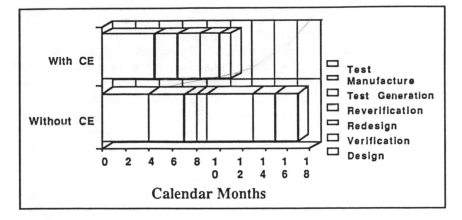

FIGURE 2-4   Time to market summary

considering the design iteration, typically amount to between 10 percent and 25 percent and result in a better quality product.

Savings in time, of course, translates directly to savings in money. Two similar printed circuit board assemblies were designed at a computer manufacturing company. Each board was made up of approximately 100 ICs, including a 32-bit microprocessor, several VLSI I/O devices, both ROM and RAM, and two ASICs.

During the first board design, there was "no time" for concurrent engineering. One of the ASICs had to be redesigned because there was "no time" for board level design verification prior to committing to a hardware prototype (which did not work). The first experience was painful and expensive. A decision was made to do the second board "right the first time." The result was a $50,000 savings on the second design when compared to the first design.

One of the most significant cost factors in new electronic device and printed circuit board designs is the cost of (and therefore time required for) high fault coverage test program generation. Many sources quote test as a 35-percent to 45-percent contributor to total product cost, and some have put forth numbers as high as 70 percent.

Since concurrent engineering includes the design for testability principles, it can help to reduce the time and cost of test generation, while simultaneously helping to increase fault coverage (i.e., test quality, which relates directly to product quality). Reductions of up to 50 percent in test program generation and fault simulation times, while still achieving 99.9+ percent (i.e., 1,000 PPM defect levels) fault coverage levels, are quite typical.

Service costs can also be reduced in several ways. The cost of a field service call continues to rise due to increased product complexity, increased personnel costs, increased spares inventory costs, increased travel expenses, and heightened customer expectations. If systems can be diagnosed remotely, boards (instead of people with boards) can be dispatched for customer replacement. Proper design for serviceability, as part of the concurrent engineering discipline, can significantly cut service costs.

NCR Worldwide Service, for example, actually supplies NCR manufacturing with funds for service connectors and EEPROMs that are put on certain products. The savings in service costs more than pays for the added parts costs (which, since they are paid for by the service organization, do not impact the accounting department's interpretation of "manufacturing costs"). There are also many more creative ways to save time and money in areas other than manufacturing, test, and service. Shortened cycle time, for example, can help to reduce inventory levels, thus saving interest costs and freeing up working capital.

The bottom line, then, is that the proper application of concurrent engineering can increase profits and make an organization more competitive. Implementing concurrent engineering is not easy and cannot be done instantly. But it can be done. Yes, it takes investment—nothing comes for free! Yes, it takes commitment—nothing happens overnight! Yes, it takes culture change— the barriers must come down! It may take time and significant educational efforts to realize the full benefits of concurrent engineering. But it can, and indeed must, be done if your organization is to be competitive in the 1990s.

REVIEW

1. Product design time (a) *is* or (b) *is not* equivalent to product development time.
2. The largest portion of product cost is determined by (a) *the manufacturing process* or (b) *the product design decisions made during product development.*
3. Design decisions in a concurrent engineering environment should be made based on (a) *facts* or (b) *opinions.*
4. Product design decisions should take into account (a) *design costs,* (b) *parts costs,* (c) *assembly costs,* (d) *test costs,* (e) *quality costs,* (f) *service costs,* or (g) *all of these costs.*
5. Manufacturing, test, and service planning should be done (a) *in parallel with the design right from the beginning* or (b) *once the product design is pretty well firmed up.*

Answers: 1: b, 2: b, 3: a, 4: g, 5: a

NOTES:

3

Making Concurrent Engineering Work

There are several sets of problems to deal with, and there are no universal solutions or pat answers when it comes to making concurrent engineering work. There are problems and conflicts to overcome in many parts of the organization and with many of the elements that are supposed to support the concurrent engineering environment.

In this chapter, we'll look at each of the most common problems that represent barriers to concurrent engineering and explore some of the possible solutions to breaking down those barriers.

ORGANIZATIONAL CONSIDERATIONS

The traditional functional organization of a company is illustrated in Figure 3-1. What stands out immediately is the separation, rather than integration, of the various functions and disciplines in the organization. This kind of an organization promotes separatism, empire building, and barriers to communications. It also makes interacting on technical matters difficult, as that interaction must often be "filtered" through a layer (or two) of management. That filtering process can induce inaccuracies and *it takes time.* The structure shown in Figure 3-1 is not conducive to a successful concurrent engineering environment.

The concurrent engineering organization, on the other hand, has contributors from each discipline reporting to a product (or project, or business unit, etc.) team leader (or manager). As illustrated in Figure 3-2, team members interact directly with each other without intervening filters. Conflicts that cannot be resolved by the team according to the guidelines presented previously can be decided by the team leader.

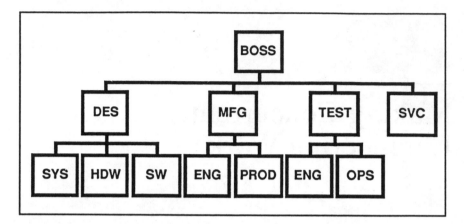

FIGURE 3-1   The traditional organization

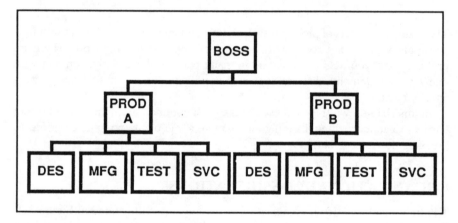

FIGURE 3-2   The concurrent engineering organization

It is also possible to have a "matrix" organization, where members of each functional discipline report directly to a functional boss (as illustrated in Figure 3-1), but are assigned to product teams (as illustrated in Figure 3-2). But matrix organizations present their own problems in terms of "Who really is the boss?"

The structure shown in Figure 3-2, even if implemented with matrixes, is the structure needed for successful concurrent engineering efforts.

At H-P/Apollo, for example, design of the DN3040 workstation followed a pattern that had all of the critical elements—marketing, design, manufacturing engineering, test engineering, industrial design, materials, quality, service, and even documentation—involved almost from the beginning of the project. The first four functions just mentioned were involved in the concept phase of the product development. The others were involved during design refinement—well before a prototype was created. Early involvement of the right resources at the right time make it possible to eliminate design characteristics that make products more expensive to manufacture, test, and service than they need to be.

Given tight development schedules, many products that are developed sequentially are more expensive to produce than they have to be. The fundamental disadvantage of the serial design approach is the cost of changing the initial design. As illustrated earlier in this text, the later in the development cycle that a change is made, the more expensive it is—usually by an order of magnitude. Thus, even if the formal organization chart is not changed, the way the organization works must change if the advantages of concurrent engineering are to be achieved.

Managers of concurrent engineering teams must understand that making concurrent engineering work involves both the implementation of technical guidelines and constant communication between all of the members of the product birthing team and their manager. But there is often a lack of *real communication*. If you are a manager, your engineers may often think that you don't support their efforts to improve things. You may think you've given them the responsibility to do it on their own and are supporting them by having paid to send them to a Concurrent Engineering Seminar™ and/or a Concurrent Engineering Technical Session™ course. You may also think that, by having educated yourself on the topic of concurrent engineering, you are now going to delegate the concurrent engineering task. You are *managing*. You expect others to *do* the "concurrent engineering."

Concurrent engineering, however, is just like quality. It takes commitment from the top and constant reinforcement of that commitment to all of the team members. It takes education at all levels. But concurrent engineering is newer than quality and not as well understood. So take the leadership position and make it your job to carry the message to your people. Get your employees educated and let them know that you are truly committed to concurrent engineering. If you don't, just like factory workers who best know how to improve productivity, but don't feel that management cares to hear their suggestions, your engineers will continue to act as they do now and nothing will change.

ACCOUNTING CONSIDERATIONS

The traditional accounting systems in most companies are an impediment to the concurrent engineering process. In fact, they sometimes seem to be "black holes" into which reams of data disappear each month, never to be seen again! Each department's cost data is measured individually (see Figure 3-3) and reported on without regard to the impact of *investments in the company's performance as a whole*. If design engineering goes over budget a few percent to make sure that

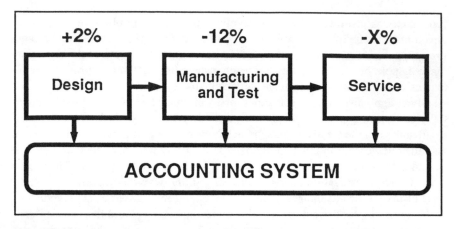

FIGURE 3-3   The traditional accounting system

everything can be built, tested, and serviced at significant savings, they tend to be penalized instead of rewarded. What kind of a system is that? Also, there is no way to "reward" the design function by transferring savings back to it.

It seems sometimes that the interrelationship between functional organizations (organized in the traditional manner) is invisible to the accounting system. The accounting system should measure product profitability, from design through manufacturing, test, and service, on a product-by-product basis. Then, realistic comparisons can be made, and cost data supporting further concurrent engineering efforts can be gathered.

How do we solve this dilemma? Here are some suggestions:

- Find out what things *really* cost
    - By operation
    - By product
- Develop alternative methods
    - Spread sheets short term
    - Activity based costing long term

Concurrent engineering involves quantitative trade-offs over the life cycle of a product. If product birthing teams have inadequate or erroneous data, they will make less-than-optimum decisions. Thus, one of the first steps in implementing a concurrent engineering culture is to find out what things really cost, no matter how painful or embarrassing that may be to some elements of the organization.

Most of the major specific cost elements, both by operation and by product (or project) are detailed in Chapter 2. You can use spread sheets in the short term, supported by flow charts and formulas (see Chapter 10), to create your own equivalent of activity-based accounting methods. For the long term, either modify the company's existing system to support concurrent engineering or convert it entirely.

The management level concurrent engineering accounting model is illustrated in Figure 3-4. Design costs for each component, subassembly, and system need to be estimated. "Deltas" required to add resources for concurrent engineering must be identified.

Tooling costs for various fabrication and test alternatives need to be determined. Material costs are also important, but are not the overriding factor in total product costs. Trade-offs and reductions in assembly, test, scrap, and service costs, not to mention getting to market sooner, often make it very advisable to add a few percent to design costs and to material costs to achieve the optimum balance.

You can create these models yourself for an individual team's use, or you can contract with outside consultants to have the necessary models created. The key point is that no single operation can be looked at in a vacuum—they are all interrelated.

There are some additional things that you can do to help solve the accounting problems that could inhibit the optimal implementation of concurrent engineering. These things include:

- Providing activity-based cost data to design teams to facilitate quantitative design
- Speeding the flow of data back to design teams
- Using accounting for process control in addition to historical reporting
- Practicing "open book" accounting

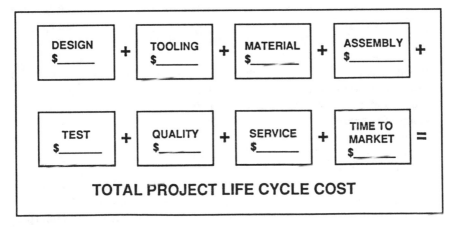

FIGURE 3-4   The concurrent engineering accounting model

As was briefly mentioned earlier, accounting data—or more accurately, cost data—is of no use to the product development teams if they have no access to it. Cost data should not be a secret that is closely guarded by the accounting department. It should be formatted for development team use and provided to the teams as quickly as possible.

In a manufacturing environment, quality, yield and defect (variance) quantity, and distribution data can be used to improve the manufacturing operations. The same is true of using cost data supplied by accounting—and by purchasing, for new proposed buys for new designs—to improve both the design process and the operation of the overall business. The books should be open to those who need the information contained in them, to make informed decisions.

PHYSICAL LOCATION CONSIDERATIONS

There is also, of course, the problem of different physical locations for the different functional groups within most companies. The managers may be on "mahogany row." The design engineers may be upstairs in carpeted offices with windows. The manufacturing, test, and quality engineers could be on (or adjacent to) the manufacturing floor. The service group is probably off in a corner (or out in the field). Nobody knows where marketing and purchasing are, and some are surprised to find out that there are people in the company who understand safety, reliability, compliance to international standards, and so forth.

There are several solutions to the physical location problem. Some require capital investment; others don't. One way to improve communication between members of a product development team in the concurrent engineering environment is to locate the members of the team in close physical proximity. This facilitates communication on an ongoing basis (i.e., between formal meetings such as presentations to management or design reviews) and fosters a group spirit of responsibility for seeing that the product gets to market as quickly as possible with the least amount of trouble.

If concurrent engineering is implemented in a matrix manner, and if people representing the various disciplines spend only part of the time in the project team's physical location, provide work spaces for the occasionally called upon team members. The presence of their space will remind the other team members not to forget them (or to get their input).

If it is not possible to create close physical proximity between members of the product development team, it may be possible to create close electronic communications between them through the installation of a network or through the use of commercially available E-Mail services, as illustrated in Figure 3-5.

The ideal situation, of course, is a company-wide network that uses a transparent user interface to a unified data base. In the face of a lack of tools to implement

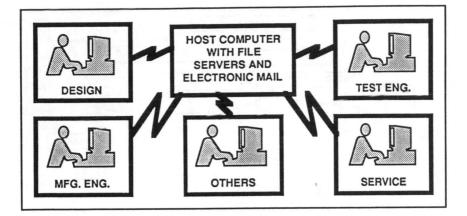

FIGURE 3-5   Using networks to overcome physical location differences

the ideal situation, any implementation that speeds two way communication between team members is a step in the right direction. If a network or the use of a commercially available network is not currently feasible, resort to "foot net" or "mail net"—transferring diskettes or modeming files at the right times to get input from each team member.

Another alternative, or sometimes adjunct, is to move the product development team with the product as it progresses through the various design phases. This approach has the advantage of giving all of the members of the product birthing team firsthand experience of the impact of the decisions that they have previously made.

This approach also facilitates continued employee education and involvement and builds an experience base that can be called upon in the future to minimize the amount of outside expertise needed on each product development team.

Team members will know which designs were easy to build, test, and service and will be in a much better position to reuse past designs (which further shortens time to market and reduces product development costs).

DESIGN ENVIRONMENT CONSIDERATIONS

What is really needed to fully automate the implementation of concurrent engineering in a transparent environment is a system like the one shown in Figure 3-6—a single design representation, a common design language, a transparent framework, tools that all work together with the same interface, tools that work from chip level to board level to system level, and expert systems to make sure that we have not violated any design (for any "-ility") rules.

What we have today are "islands" of design, manufacturing, and test automa-

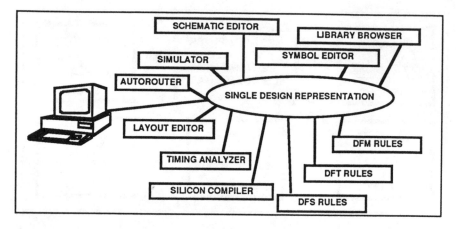

FIGURE 3-6   The ideal design environment for concurrent engineering

tion, some linked together well, some not so well. But progress is being made, even if too much emphasis is still being placed on "front-end" design tasks by most CAE vendors. Work is progressing on open system architectures, design frameworks, and inter-tool communication (although mostly at the application specific device level, with board and system level tools still lagging significantly).

Until we have the ultimate computer-aided concurrent engineering "tool set," the communication and the implementation must be done via a very low technology medium—*human action and interaction.*

If we look at the tools available for concurrent engineering as of this writing, we see a list like the following:

- ASIC level design and simulation
- Board level design and simulation
- PCB design CAD
- ATE systems
- "Linking" software from some design platforms to some test platforms at some levels of design integration

There are specific tools for each specific level of design integration. There are specific tools to perform specific tasks at each level of integration, and some of them run on common platforms. There are also some links between the hierarchical levels and some links between the tasks. What's needed, in terms of CAE support of concurrent engineering, however, is *not links*. It is *integration!*

Listed below are some of the design automation items necessary for the automated, rather than manual, implementation of the concurrent engineering discipline. It is encouraging to see articles in the technical and business press that development of at least some of the items on the list is beginning.

- CAE frameworks that support *all* engineering tasks
- Library elements than can be reused at various levels of integration
- Common languages and formats for design information
- Common data bases and file servers at all integration levels
- Synthesis tools that include CE expert system rule checking

Professional organizations like the IEEE are also getting involved in trying to standardize many of the above items on frameworks that support open architectures. But until these tools are actually developed, debugged, and deployed in most organizations, organizations practicing concurrent engineering will have to work with the tools available while encouraging their design automation suppliers to develop the items listed above.

Concurrent engineering involves the simultaneous design of both products and processes. One of the best ways to get designs and processes right the first time is to simulate them. This helps to reduce design iterations (which add to both development cost and time to market) and results in better quality product and process designs.

Most of today's tools simulate either product or process, but not both. This will change in the future, however, to allow for simultaneous simulation and for networking design data bases to manufacturing process and control data bases. Some of this has already happened in the mechanical design areas. It will eventually happen in electronic and other design areas as well.

Integrating the electronic design automation environment is not a trivial nor inexpensive task. Successful integration requires significant levels of support. Mentor Graphics, for example, points out the need for administrators, developers, and librarians in support functions so that the product design teams can concentrate on product designs.

The use of centralized support groups also minimizes duplication of effort (and filling up disk drives and file servers with redundant information). As in all support or staff functions, care must be taken to manage the support groups so that they remain the means to facilitate concurrent engineering (and not the limiting or controlling force).

A recent (1990) survey taken by the Adams Co., Palo Alto, CA, of readers of Electronic Design magazine, listed the tasks that the readers said they would perform more often if they were supplied with better and more comprehensive design automation tools. A summary of that survey appears in Figure 3-7. Four of the top five tasks involve some form of simulation—an attempt to make sure that a design works right the first time. The other task mentioned in the top five is test generation.

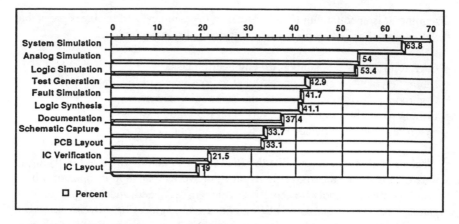

FIGURE 3-7   Priorities for better CAE tool development

This data suggests two things. The first is that people want to do their jobs right the first time. The second is that those whose job it is to select new design automation hardware and software systems need to make sure that future investments in design automation are purchased in the right order. The only way to get design automation vendors to supply the tools needed to help automate the concurrent engineering process is to apply economic pressure on them. Who usually supplies economic pressure? (Answer: the customer!)

GOALS, STRATEGIES, AND TACTICS

The goals for an organization are usually set by the senior managers of the company. It is also the senior managers who promulgate the concurrent engineering mission statement and set the goals for the product development teams. Care must be taken, however, to empower the team leaders and team members so that they can develop their own strategies and tactics for accomplishing the required goals. Team leaders should be told what needs to be accomplished, and why, and then allowed to get on with the task. If senior management second guesses the product development teams, or arbitrarily dictates design approaches, the process will not work.

A very prevalent lament from engineering teams is that the many of the goals handed down by upper management are unrealistic and have been dictated to the team. This fosters passive-aggressive behavior that leads to poor morale and consistently missed development schedules.

For the concurrent engineering process to work, team members must agree with the goals and must agree with the schedule. Senior management must accept that sometimes things take longer to do than it would like, especially if things are going to be done right the first time.

Team members will be very unwilling to be held to goals or schedules handed down from on high that they feel are not doable. Once team members have agreed to the goals and schedules, however, they can be made accountable for both their own individual tasks and for the cost and schedule performance of the team as a whole.

There are some well-tested and well-proven guidelines for team building that fully apply to the development of product birthing teams in the concurrent engineering environment. The guidelines, which apply to both senior managers and to the product birthing team leaders, may be summarized as follows:

* Tell what result is needed, not how to accomplish it
* "Hire" to strengthen weaknesses
* Help avoid mistakes already made
* Give credit where credit is due
* Reward supporters as well as superstars
* Interface with others regularly
* Lead by example

When you think about them, these guidelines (or even rules) are pretty much common sense. But common sense is not always a common commodity, especially in the heat of trying to get a design completed under heavy schedule pressure. So ignoring any of the rules listed above is inviting less-than-optimum team performance (and a resulting less-than-optimum product design).

In particular, there are many "war stories" in each organization about design attributes that caused considerable pain in either manufacturing, test, or service (or all three). Team members should share these tales so that similar occurrences can be minimized or eliminated. Don't wait for formal meetings to communicate. Make constant communication a requirement.

Another consideration is how large the product development team should be. The results of a survey by EE Times magazine (October 15, 1990) yielded the interesting information shown in Figure 3-8. It is clear that teams are much larger in Japan than they are in the United States. It's also clear that many U.S. engineers feel underutilized. LSTI's experience in conducting seminars and workshops confirms this feeling, especially for engineers involved with manufacturing and test. They would like to be far more utilized on new project design teams.

The optimum team size and membership will, of course, vary from company to company and from product to product. The point is not to shortchange the team by excluding anyone with expertise to contribute. The more up-front input, the less chance of a later "gotcha" that could either cause a design iteration or result in a product that is more expensive to produce or service than it should be.

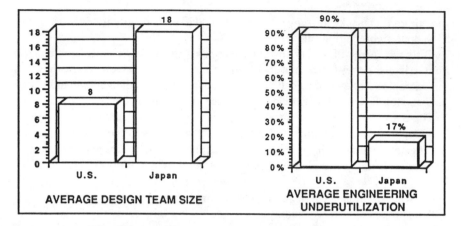

FIGURE 3-8   Concurrent engineering team size

Finally, it may be necessary to change the company's performance appraisal system to help engineer the culture change to a concurrent engineering environment. That means getting input not just directly from the team members' direct supervisors (in a functional management organization where people are allocated to the various teams), but also from the project team manager and even the other members of the project team. When people are aware that they will be judged on team performance as a major part of their performance appraisal, they take team success much more seriously.

Team members will also be more inclined to do their best for the team when progress or design review presentations must be made to senior management. Team leaders will want to show a cohesive, comprehensive design approach, rather than having team divisiveness surface during such meetings.

HANDLING CONFLICTS IN THE CONCURRENT ENGINEERING CULTURE

There will inevitably be conflicts between the requirements and desires of each member of the product birthing team. That's not only to be expected, it is healthy— if channeled properly. The guidelines for handling conflicts in the concurrent engineering environment are:

- Remember the customer's needs
  - External
  - Internal
- Listen to each team member's input
- Find out *why* there is a conflict

- Engineer a compromise
  - "What if *we*..."
- Exercise consensus management

Customer needs (external) come first—by a factor of ten. Your company's customers pay your salary. Internal customer needs come next. Every person's point of view deserves a fair hearing. Try to really listen (rather than thinking up a rebuttal) to each input. Make sure you understand what was said by paraphrasing it back to the speaker (i.e., "I heard you say '...' "). Use the word "Why?" over and over again to get to the root cause of a conflict. Once the root cause is identified, a compromise can often be engineered that satisfies both parties.

Since one of the basic tenets of concurrent engineering is quantitative design, the concurrent engineering cost model can often be used to take the opinions out of technical or strategy disagreements between team members. You can play "what if..." with alternative number one, alternative number two, and any compromise alternatives that the product birthing team can come up with. Then simply calculate the impact of those decisions on the overall business revenues and costs over the life of the product. The one that maximizes overall profits, taking into account the impact of time to market, is the choice—every time.

SUMMARY

Making concurrent engineering work requires management skills, realism, and leadership. It also requires commitment on the part of both management and the members of the product development teams. Organizational structures need to be modified to support the concurrent engineering process, and activity-based costing systems need to be set up to support the efforts of each of the teams within the company.

Performance appraisal systems may need to be changed in order to better track individual contributions to team cost and schedule efforts. New CAE and networking tools can often be used to help overcome physical location barriers, but existing methods—including phone calls, faxes, and personal meetings—can be used to implement concurrent engineering without the requirement for large capital expenditures for new tools. Everyone in the organization must, however, be fully trained in concurrent engineering, speaking the same language and pursuing a common set of goals that they have agreed to.

REVIEW

1. The traditional functional organization (a) *helps* or (b) *hinders* the implementation of concurrent engineering.

2. Current accounting systems provide (a) *good data upon which to base design decisions* or (b) *historical data the leads to punishment more often than improvement.*

3. It (a) *is* or (b) *is not* necessary for all team members in a concurrent engineering process environment to be physically located together.

4. Concurrent engineering (a) *will* or (b) *will not* work if senior management dictates schedules to teams that cannot be met.

5. Conflict in a concurrent engineering environment (a) *is* or (b) *is not* healthy and desirable.

Answers: 1: b, 2: b, 3: b, 4: b, 5: a

NOTES:

4

Managing the Change to Concurrent Engineering

Concurrent engineering is an ongoing process, not an instant fix program. It is also a continuous improvement process, very much like total quality management (TQM). Managing the change to concurrent engineering takes several things, among them the following:

- Top-down specifications
- Bottom-up design activities
- Management involved and committed
- Everyone educated and speaking a common language

Creating top-down specifications will get easier the more often people do it and see the benefits of doing it. Designers will implement the concurrent engineering guidelines more often when they realize that it doesn't have to take a lot of time or cost a lot of money, and that doing so will shorten time to market and simplify their own design verification tasks.

Management must be both committed and involved. If you are a manager, you must carry the concurrent engineering message to the team leaders and team members in your organization. You can do it. You have that power. Your people need education. Let them know that in a nice way, educate them formally yourself, or arrange to have someone bring a complete Concurrent Engineering Program™ in-house, tailored to meet your exact needs.

Remember to communicate regularly with both your internal and external customers, for they pay your salary. Lastly, recognize that *implementing the concurrent engineering process in your organization takes commitment.* Without it, there will be no results.

PHASES ON THE PATH TO CONCURRENT ENGINEERING

The evolution of an organization from a serial (or sequential) design culture to a concurrent design culture is very similar in many ways to the evolution of an organization toward total quality management. In Sematech's Partnering For Total Quality series, the five specific phases listed below are identified:

- Short-term focus
- Product focus
- Product and service focus
- Process or system focus
- Continuous improvement focus

Many companies today have already moved to the third phase mentioned—the product and service focus. The objective of concurrent engineering management is to help you move to the fifth phase—a continuous improvement focus. In the next few sections, we'll look at each phase, adapted from the Sematech material for concurrent engineering, in a little more detail.

Phase One—The Short-term Focus

The short-term focus phase is the unfortunate phase that too many companies still find themselves in today. It may be characterized as follows:

- Revenues and budgets are a higher priority than concurrent engineering.
- No concurrent engineering mission statement exists.
- Little or no data is available or used.
- Intuitive design practices are used.
- A high incidence of scrap and rework exists.
- The sales department is the only customer contact.
- Only skill-related, on-the-job training is provided.

Organizations in this phase have no mission statement and no management vision beyond the short-term "bottom line." R&D is often starved to shore up quarterly profit statements, mortgaging the company's future. Intuitive design practices are used, so every product design looks different and has its own manufacturing, test, and service idiosyncrasies.

Little or no manufacturing/quality data exists, and, if it does, it is not analyzed nor utilized properly for process improvement. Scrap and rework activities make up a large portion of the manufacturing burden, often representing "factories within factories" that deal with up to 30-percent defective products. Minimal feedback and cooperation exists between manufacturing and design engineering, and designers are insulated from customers.

The emphasis on budgets above all else means that professional training of employees is minimal to nonexistent. The managers of these companies are too busy fighting forest fires to practice fire prevention by eliminating the arsonists.

Phase Two—The Product Focus

Companies that have recognized the peril of the short-term focus mentality have moved toward a product focus mentality in an attempt to regain lost market share and become competitive again. This phase may be characterized as follows:

- Design quality is viewed as meeting specifications.
- Design mission does not include DFM, DFT, and DFS.
- Data is applied to factory improvements only.
- Consistent design methods are lacking.
- Moderate to high scrap and rework costs exist.
- Senior executives meet only key customers.
- Development personnel are insulated from customers.
- Training is limited to skills-oriented training.

Consistent design methods are still lacking, and design quality is viewed simply as meeting customer specifications (rather than meeting or, even better, exceeding customer *expectations*). "Quality data" is beginning to be gathered and analyzed, but this is applied only to the manufacturing operations in an attempt to "lower costs" by reducing rework and scrap costs, many of which are aggravated by product design characteristics. The topics of Design For Manufacturability (DFM), Design For Test (DFT), and Design For Service (DFS) become major topics of conversation in the operations side of the house, but are ignored by the development side.

Senior managers are beginning to see the value in getting "closer to the customer" but are still too busy fighting fires to meet with any but the key customers. Development engineers get customer input only through the sales/marketing staff and the occasional "blast" from top management as a result of a customer visit where company and product deficiencies were painfully pointed out. Does this sound at all familiar?

Phase Three—The Product and Service Focus

Because of the beating that many electronics companies took in the 1980s from foreign competitors on the time to market, cost, and quality fronts, many companies have now moved into phase three—the product and service focus. Phase three can be identified in an organization by looking at the following characteristics:

- Design quality is viewed as including DFM, DFT, and DFS.
- Concurrent engineering mission is product focused and is an executive priority.
- Data collection and analysis activities are in place and functioning.
- Consistent design methods are used for products.
- Lower scrap and rework occurs.
- Periodic customer surveys determine expectations.
- Some concurrent engineering teams exist.
- Training is available for all affected employees.

Much more enlightened managers recognize the need for concurrent DFM, DFT, and DFS, and those activities are at least minimally supported by data collection and analysis activities that are in place and functioning in the factory in support of statistical process control and total (factory) quality management efforts. Guidelines and checklists exist to help insure some design consistency from product to product.

As a result of these investments in both money and people, these companies are seeing reduced scrap and rework costs. The continuation of DFM, DFT, and DFS efforts that are making that possible is an executive priority and some concurrent engineering teams exist. This is usually a result of professional training for the employees most affected by design decisions (i.e., everyone other than the designers) and their subsequent efforts to improve things.

Phase Four—The Process or System Focus

The fourth phase, which some companies have now moved into, is the process or system focus phase. Senior managers have recognized the value of cross-functional design teams and, as a result of the vision and efforts of typically one senior manager, concurrent design methods are used for all products. Phase four characteristics include:

- A senior executive owns the concurrent engineering mission.
- Cross functional concurrent engineering teams are functioning.
- Widespread internal and some external quantitative data exist.
- Concurrent design methods are used for all products.
- Manufacturing, test, and service costs are reduced.
- Time to market for new products is reduced.

Widespread internal and some external quantitative data on product design requirements, customer designed features, and product quality results exist, and statistical process control is used extensively in the factory and is beginning to be used to improve the design process. Overall business costs are being reduced, and time to market for new products is being significantly shortened.

This is the phase that many of the companies used as examples in Chapter 1 of this text are in, and it is a significant improvement over the previous three phases. Getting to this phase took significant investments in training, both organizational and technical.

Phase Five—The Continuous Improvement Focus

The continuous improvement phase of concurrent engineering is the "world class" phase. The characteristics of an organization that has evolved to this advanced state are as follows:

- Employees are completely empowered to fulfill the company's concurrent engineering mission.
- Real time statistically valid internal and external quantitative design and customer requirements data is used.
- All affected employees are members of concurrent engineering teams.
- Expanded partnering exists with both customers and suppliers.
- Design methods are benchmarked against the best competitors.
- The positive impact of training programs has been proven.

This phase is where products can be customized rapidly to meet more frequent customer requirements changes through the reuse of engineering investments, where order and factory cycle times have been cut significantly, up-front planning has replaced back-end redesign, reaction, and rework, first pass factory yields are in the high-95+ percent ranges, and people are motivated and proud of their jobs.

This phase is also where your organization must be to compete successfully in the 1990s. In addition to the obvious benefits of concurrent engineering that were pointed out during the introduction chapter of this text, some subtle benefits (listed below) also come into play:

- Lower capital equipment cost
- Greater use of automation
- Less chance of redesign
- Fewer parts to buy from fewer vendors
- Better factory availability
- Improved design quality
- Improved organizational motivation and morale
  These subtle benefits position a company to respond ever more quickly to

changes in the marketplace and to take rapid advantage of new technologies—both product and process.

ENGINEERING CULTURE CHANGE

Make no mistake about it. The shift to a concurrent engineering environment is a culture change. And that scares a lot of people. That's reality. But recognize reality as what is—it is the way things are. It may be right, it may be wrong. It may be comfortable, it may not be comfortable. But it is what it is.

To begin to engineer a culture change, it is critical that you acknowledge the realities of the people, the products, the organization, and the politics around you. Realities such as which people are in the real power positions in the organization, what would make them want to help you effect the shift to concurrent engineering and what you need to do to as a top manager to persuade them—not just order them—to your point of view.

But don't be afraid to recognize that *reality is something that you can change.* Also recognize that things will stay the way they are without positive action on your part. Then ask yourself the question: *Do you want things to stay the way they are?* If your organization is in phase five and you answered "yes" to the preceding question, great. If it is not and you answered "yes" to the preceding question, then you've already spent far too much time reading this book. You should really be looking for another job in another industry.

The primary guidelines for engineering culture change are illustrated in Figure 4-1. They begin by centering on the concept of open book management.

Empowering employees means giving them the information they need to do

---

Concurrent Engineering Management ━━━━━━━━━━━━

Open Book Management

- **Demystify Strategy**
- **Track Times And Costs**
- **Report Product Profitability**
- **Explain Company Performance**
- **Report Division Revenues**

FIGURE 4-1   Guidelines for implementing culture change

their jobs without constantly preparing infinite numbers of justifications and referring them to upper management for approval. Open book management does just that. It lets information permeate a company.

INC. magazine (September 1990 issue) had an excellent article on the subject, and it is recommended reading for all readers. People basically want to do a good job, but they often don't know what a good job is or how their current performance is contributing to (or taking away from) company results. Open book management gives them insight into the effects of their actions. It can also relieve significant burdens from top management by driving decision making down to the lowest levels. Reporting product profitability, company performance, and division revenues (where applicable) can foster both internal (healthy) competition and company-wide cooperation.

Listed below are seven guidelines for effecting the changes needed in your organization to move from the old serial engineering method to the new concurrent engineering method.

- Understand your existing culture.
- Encourage those who want to change it.
- Use the best "subculture" as an example.
- Don't attack the current culture head on.
- Don't count on a vision to work miracles.
- Give things time to change.
- *Live* the culture *you* want.

Take a look at the reality of your existing culture. Then make a mental picture of how it *should* be. Find those in the organization who can share your vision. There are usually "pockets" of forward thinkers in any company. Use them as an example to effect gradual change. Attacking a culture head on is not recommended—there is too much history and too much inertia. And while your vision, promulgated through the concurrent engineering mission statement, will help you make the changes needed eventually, don't count on instant miracles from the organization. You, however, can begin living the new culture today. You have that power.

It has also been well documented in the current business literature that a "bias toward action" is a characteristic of an "excellent" organization. What hasn't been explicitly stated is that excellent organizations are made up of excellent people. And excellent people have a bias toward action. They seek constant improvement in designs and processes. They try new things. They adapt to new circumstances. They avoid "paralysis by analysis." They know that momentum is important and that excellent companies adapt. They also know that any decision is (usually!) better than no decision at all. Does this make sense? Try it you may like it!

You have a considerable body of new knowledge as a result of investing your time in reading this book. Will you act on that knowledge? Will you use the techniques that have been presented? Will you acknowledge your own power to cause the changes that need to be made in your existing environment to be made? We hope you will. You made your first "bias toward action" decision by deciding to read this book. Don't let the new knowledge that you've gained languish on your bookshelf. Transfer that knowledge to your people yourself or have it done for you.

The implementation of concurrent engineering need not entail large capital expenditures or a complete reorganization of the company. It does entail a change in the way the people in the company work and the way in which the various functional elements of the organization interact.

The same approach that works for Total Quality Management (TQM) works for Concurrent Engineering (CE). Once the basics of concurrent engineering are in place, continue to improve the process by using the following steps:

- Look at processes as a whole.
- Engineer small changes constantly.
- Use evolutionary steps.
- Design measurements for each step in the process.
- Automate measurement and analysis as appropriate.

Overall life cycle cost can be illustrated with a "bathtub curve," as shown in Figure 4-2. The optimum point, identified by the vertical arrow, is the point we want to achieve. We can lower the build, test, and service costs by spending a lot more design time to optimize every detail of the design. But if we do that, we may find that we have indulged in overkill and missed part of our market window. We

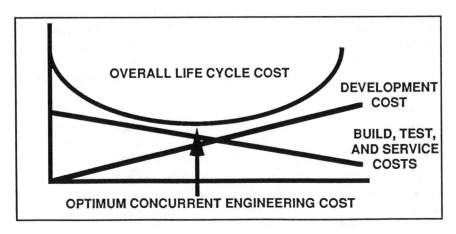

FIGURE 4-2   Overall life cycle costs

want to do enough concurrent engineering to balance the design efforts with the production and service costs. The two slanted lines in Figure 4-2 actually intersect past the optimum cost point. Know when to say "enough is enough, let's run with the design as it is" (with the agreement of all of the members of the team).

Most products go through cycles like the one illustrated in Figure 4-3. In the early phase, sales growth is fairly rapid. In the mature phase, it levels off. As the product become obsolete (or is supplanted by products using newer technology) or as the market becomes saturated, sales fall off.

A delay in time to market can cause your company to miss part (or all) of the early fast growth phase of the market. The whole idea is to be out there first in order to gain market share during the time that the number of orders will be the largest. The implementation of concurrent engineering is designed to prevent delays and let you get to market sooner with better products that can capture more market share and contribute more profits.

CUSTOMER RETENTION VERSUS ACQUISITION

The chart in Figure 4-4 contrasts the cost of doing business with existing and new customers for existing and new products. What it says is that, given a difficulty factor of one for selling additional existing products to existing customers, it is three times more difficult to sell an existing customer a new product. That difficulty factor can be reduced if the customer feels that he has contributed to the design of the product.

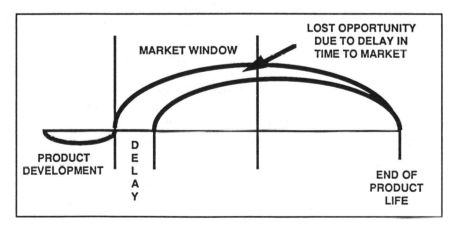

FIGURE 4-3   Market share considerations

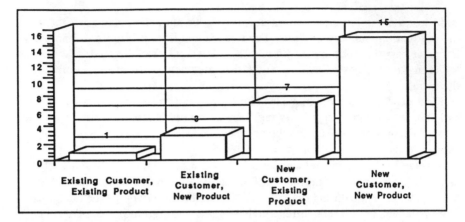

FIGURE 4-4   Customer retentions vs. acquisition

For new customers, the difficulty is even greater. Thus, when customer representatives are included, as appropriate, in the product birthing team's process and progress, they are much more likely to "own" the product even before it is completed. This again reduces the difficulty (i.e., the cost) of getting new customers (i.e., gaining market share).

USING CUSTOMER CONTACT APPROPRIATELY

Another part of staying more competitive in the 1990s is to get closer to the customer. And that doesn't just mean having the marketing and sales people closer to the customer. It means making sure that the product design team members are also in close contact with customers so that they understand exactly what it is they should be designing.

The bar graphs in Figure 4-5, also from the EE Times survey mentioned earlier, contrast the number of customers contacted by product development people in the United States and in Japan. It is clear from the data that most U.S. product development teams are still far too insulated from the customers that they are designing products for. This same situation is true for many companies in Europe as well.

Listed below are some guidelines for customer contacts with the product development teams:

- Formal customer meetings should involve teams:
    - Marketing
    - Engineering
    - Contracts.

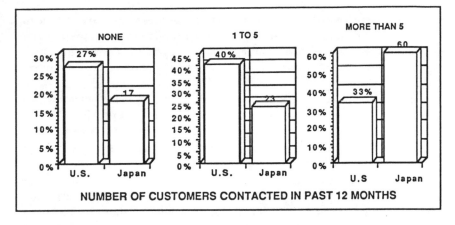

FIGURE 4-5   Customer contact comparison

- Team members should be individually accessible—Informal Questions.
- Customer directed R&D pays off.

Many marketing and sales people do not want other people talking to their customer's personnel directly. This tends to further the isolation of the design group and sets up "filters" between customers and designers that can cause misunderstandings. To be sure, there are some engineering personnel who just don't know how to act properly in front of or in meetings with customers. The solution, however, is to *train* them, not to deny them.

Customers with technical questions about products should be able to access technically competent personnel for guidance. In the process, your design personnel may gain valuable insights into what your customers may need in the future and what they think of both your company and your competition in terms of both products and services. Customer-directed R&D can pay big dividends.

USING SUPPLIER CONTACT APPROPRIATELY

Suppliers can also be valuable members of the product development team. It is almost always less expensive in the long run to buy something from someone who already makes it than it is to develop the designs, tooling, and manufacturing expertise to do it yourself. Development efforts should be spent on the value that your company adds to the completed product and not on the common components that go into it. The exception to this, of course, is in the case where proprietary component technology or processes are involved.

Partnering with suppliers adds their expertise to that of the product development team at virtually no cost. Suppliers often suggest better, less expensive ways to do things. Tying them into your technology and production needs also speeds design and facilitates just-in-time manufacturing practices. Reducing the overall number of suppliers also reduces the overhead costs of purchasing, inspection, and record keeping.

Purchasing (or contracts) people should, of course, be the official interface between your firm and your suppliers and should be the only people allowed to commit company funds. But engineer-to-engineer communication is very important in the early stages of discussion about technologies, alternatives, and design improvement suggestions.

Proprietary agreements can be put in place to protect confidential information—both yours and your suppliers—and engineers must know that any prices discussed during engineer-to-engineer discussions are not binding on either party. But prices must be at least estimated if product development teams are to make good choices.

Finally, price (or cost) is not the driver in the concurrent engineering and total quality management—oriented organization. Higher quality parts, even if the parts themselves cost more, can result in lower costs at next levels of assembly, due to increased yields. So quality and on-time delivery are more important than price alone.

THE MANAGEMENT SOLUTIONS

The first steps in moving an organization into the concurrent engineering mode are to:

- Get religion!
- Develop and promote a mission statement that all employees can understand.
- Educate the entire organization in the principles and techniques of concurrent engineering.
- Keep the process moving with constant updates and improvements.

If you want your employees to change their traditional methods of operation, you must realize that the change must begin at the top—with you. If you truly believe that concurrent engineering can shorten your development times, while simultaneously helping to improve quality and lower costs, develop a mission statement outlining your new "religion" and communicate that vision clearly and understandably to all employees.

You must commit to educating the entire organization in the principles and techniques of concurrent engineering if you expect to effect the changes needed. That education must continue on an ongoing basis to keep the process moving, to help build momentum, and to show that this is not another passing management fad. Just as quality requires top management commitment and ongoing employee education and interaction, so does concurrent engineering.

We've seen that implementing concurrent engineering requires the allocation of additional engineering resources (in excess of the design engineering functions). Where will those resources come from? If it is not possible to add resources, or if people are current spread too thinly, consider cutting the number of projects that your people are currently working on to free up the needed resources within the existing budget.

Concentrating on core technologies and taking advantage of the reusable engineering that gets created in the concurrent engineering discipline can allow you to still increase the number and variety of product offerings. Remember that concurrent engineering is not an instant fix—it will take time. This requires applying the resources required for long-term success, even if some pet projects must be eliminated in the short term.

For concurrent engineering to really work, the employees who are members of the team must have decision-making power. Trying to make them responsible for concurrent engineering, without truly empowering them with the authority to implement their recommendations and decisions, will surely lead to failure. You must stay committed and involved, making sure that everyone knows that the company is in this for the long term and that you're watching the progress that is being made.

Change the performance appraisal systems to include measurements and rewards for team performance. Team members must be responsible not only for their own portions of a project but also for the impact of their actions on the performance of the team. Use activity-based accounting to find out what things really cost and provide that data to the product birthing teams in a timely manner. "Open book" management will let employees see the effect of their actions and decisions on the business as a whole and help them recognize that they are responsible for their own job security.

THE CONCURRENT ENGINEERING COMMANDMENTS

Listed in Figure 4-6 are the concurrent engineering "commandments." Follow them and they will lead you to shorter time to market, improved quality, lower overall business costs, more market share, and increased profits.

Ignore them and your fate is in the hands of your competitors.

> • **Create Multifunctional Design Teams**
> • **Improve Communication with the Customer/User**
> • **Design Processes Concurrently with the Product**
> • **Involve Suppliers and Subcontractors Early**
> • **Simulate Product Performance**
> • **Simulate Process Performance**
> • **Integrate Technical Reviews**
> • **Incorporate "Lessons Learned"**
> • **Integrate CAE Tools with the Product Model**
> • **Continuously Improve the Design Process**

FIGURE 4-6   The concurrent engineering commandments
Note: Permission is hereby granted for unlimited reproduction of Figure 4-6 for internal use. All other rights reserved.

SUMMARY

Concurrent engineering is an ongoing process that involves changing the culture in many organizations in order to move those organizations from the short-term focus phase to the continuous improvement focus. Engineering culture changes requires commitment and constant communication—of goals at the beginning and results at the end. It also requires time and effort, as well as the ability to look at processes as a whole.

Measurements of progress must be designed, and automated after proofing, so that valid statistical data are available to support the concurrent engineering process as it evolves. Those measurements will allow you to determine the optimum engineering cost while simultaneously minimizing overall life cycle costs for new products.

Concurrent engineering managers can help to increase their company's market share by getting products to market sooner and turning new customers into existing customers who will buy additional new products with less sales and marketing effort. Customers and suppliers become partners in a concurrent engineering environment, and barriers to ongoing communications are removed. Managers must communicate their belief in the efficacy of concurrent engineering and their commitment to following the concurrent engineering commandments. They must constantly innovate with evolutionary steps in improving both product designs and the product design process itself.

Having read this far, you now have both the economic and organizational information you need to begin implementing the concurrent engineering process in your company for new designs. You can, and indeed must, make it happen. You have that power.

REVIEW

1. Successful concurrent engineering requires (a) *top down specifications,* (b) *bottom up design activities,* or (c) *both.*
2. There are (a) *two,* (b) *three,* (c) *four,* or (d) *five* phases in the path to true concurrent engineering.
3. Concurrent engineering (a) *does* or (b) *does not* require management to share performance information with employees.
4. It is (a) *harder* or (b) *easier* to sell a new product to a new customer than it is to sell a new product to an existing customer.
5. Customers and suppliers should be considered (a) *adversaries* or (b) *allies* in a concurrent engineering environment.
6. The concurrent engineering commandments specify that (a) *products,* (b) *processes,* or (c) *both* should be simulated in order to achieve a "right-the-first-time" result.

Answers: 1: c, 2: d, 3: a, 4: a, 5: b, 6: c

NOTES:

5

Design for Performance Considerations

Product size and product operating speed are two of the parameters most often cited as barriers to the inclusion of the technical guidelines for concurrent engineering contained in Chapters 7 through 9 of this text. Everyone wants everything to operate at the maximum possible speed and to perform an infinite number of functions in a minuscule amount of space. While there is nothing inherently wrong with either of those desires, they sometimes result in products that are unnecessarily overburdened with technology and complications.

There is also incessant and increasing pressure to bring new products to market faster and at the lowest possible cost. The concurrent engineering technical guidelines are designed to help you accomplish this, even if it means some minor performance trade-offs.

PRODUCT SPEED CONSIDERATIONS

One thing that appears to be a given in the design of every new product is that it must be designed to operate at the maximum clock speed possible with the very latest technology components. Speed, no doubt, is often a critical parameter for products like computers, workstations, avionics, and so forth. But is it always a critical parameter?

Before deciding that a new product design must push the operating speed requirements to the maximum available, ask yourself the question "Why?" a few times. You may find that your answers conform to the "80/20 rules." That is, 80 percent of the time there is no reason why you have to maximize the speed of every path in a circuit design, while 20 percent of the time you will probably have to. The key is in prioritizing the speed issue. Does a keyboard need a 100 MHz clock speed to interface to a human being?

Some of the questions that must be answered regarding the product operating speed issue include the following:

- What is the application environment?
  - Real Time
  - Background
- What speed is actually required?
- What minimum technology will meet the requirements?
- Are price/performance issues fully understood?

Real time systems, for example, must typically operate faster than systems performing background (or occasional) operations. What speed is really needed?

Older, proven, more mature, and better understood components and technologies can often provide the required performance in new designs. The parts are usually available from multiple sources and the prices are lower. Why use a RISC processor for an answering machine when a small microcontroller chip will do the job? Why develop an application-specific integrated circuit (ASIC) when there is room for lots of discrete "glue logic" devices?

In a concurrent engineering environment, we try to avoid the tendency to use new technology just for the sake of using it (or learning it or playing with it!). Make sure there is a legitimate business reason for each decision made regarding the questions above.

PRODUCT SIZE CONSIDERATIONS

Another common performance issue is the "space squeeze." There are often legitimate restrictions on the size that a product must be. Building a new subassembly to fit into an existing space in an existing system is a prime example of that. So is designing a product that must fit into a standard size shirt pocket.

At other times, however, size (in the plus or minus 5 percent range) is not such a critical parameter. Ask "why" the product must be the size it is. If you ask often enough, you may find that the answer is not really valid and that product size could be increased marginally to improve its manufacturability, testability, and serviceability. There are even cases where making something bigger will improve customer perceptions and acceptance of the product!

The drawings in Figure 5-1 illustrate the points concerning product size considerations.

In the example on the left in Figure 5-1, it is painfully obvious that a very concrete space constraint exists. Thus, implementing the concurrent engineering

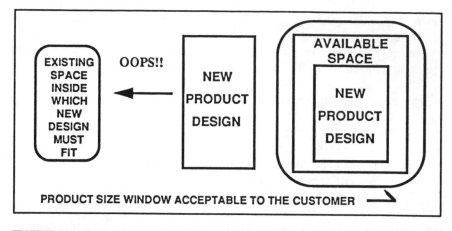

FIGURE 5-1   Product size considerations

guidelines will take some additional design creativity—the same kind of creativity required to fit all the required functions into the available space.

In the example on the right in Figure 5-1, however, there is considerably more flexibility in product size decisions. As long as the product fits within the size window acceptable to the customer, it can range in size from the smallest possible new product design to a larger one that can accommodate almost any of the concurrent engineering guidelines requested by the manufacturing, test, quality, and service engineering organizations.

PRODUCT WEIGHT CONSIDERATIONS

A similar set of trade-offs must be considered where product weight is the issue and much depends on the function of the product and its intended use. Portable computers (including "notebook" models) must be as light as possible. The lighter they are, the better consumers like them. Avionics systems are another example where weight is a critical performance element. Telephones, on the other hand, must often be made heavier so that they don't slide around when the handset cord is stretched.

Actually implementing the concurrent engineering technical guidelines, however, usually has minimal, if any, impact on product weight.

HUMAN FACTORS CONSIDERATIONS

Another critical set of design factors are the human factors, or ergonomics, that are considered and incorporated into the design of a new product. The following list is only a partial list of all the things that need to be considered, but it is representative of the categories that are important:

- Is operation simple and foolproof?
- Have all safety factors been considered?
- Is user assembly/disassembly required?
- Is service assembly/disassembly required?
- Is the product aesthetically pleasing?
- Does the product match its intended environment?

What we're really talking about here is paying attention to details. One of the biggest criticisms of new products is that they don't always "look right" or "feel right." Thus, the marketing organization should have a significant input (but not an absolute input) on product packaging design, front panel layout, the location of interface connectors, the serviceability requirements, the intended user environment, and so on.

Let's all face reality: A product that does not perform a useful function is a useless product that will not sell. Close relationships with customers are critical to identifying the functions that are really required in a product. Adding "bells and whistles" just because they can be added can actually hurt a product. The customer may perceive it not as "more powerful" but as "more complex"!

A product obviously has to perform all customer required functions. Adding additional *useful* functions can give a product a marketing edge. Adding functions just for the sake of adding them can make the product more difficult to produce, verify, deliver, and service. It can also delay the introduction of a product, which is counter to the aims of concurrent engineering. We want to decide up front what features are really needed in order to prevent the "moving target" syndrome.

Other performance factors, such as speed of operation and reliability, are also important. Remember that human beings use most of the products that we design—make them simple and safe to operate for the end user and simple and safe to produce in the factory and service in the field.

QUALITY CONSIDERATIONS

The basics of meeting customer quality expectations are well documented, and the techniques for their implementation are the subject of many seminars and programs, including those by Crosby, Deming, and others. Concurrent engineering is actually just an extension of design for quality where we look not only at the "external customer," or end user of the product we design, but also at the "internal customer"—the people and organizations to which we provide our design "product."

High quality is a design requirement. Products without quality designed in will eventually fail in the marketplace. They may even fail to reach the marketplace. And what good is a design that never sees the light of successful implementation and use? In a concurrent engineering environment, quality engineers are part of the product birthing team right from the start of a project.

PRODUCT COMPLIANCE CONSIDERATIONS

Another area to be considered is that of compliance with the various regulatory agencies in each of the countries around the world. Thus, team members with expertise in things like the FCC (Federal Communications Commission), UL (Underwriters Laboratory), CSA (Canadian Standards Authority), and other regulatory agencies may also be required as part of the concurrent engineering effort.

It is sad, indeed, to find out that a product won't meet regulatory requirements after the design is complete. This may entail added costs, for example, for extra shielding (for RF emissions) or a complete change in packaging materials or methods. That kind of a retrofit design iteration can really impact time to market. If engineers with compliance expertise are part of the product development team, iterations can be reduced or eliminated.

PRODUCT RELIABILITY CONSIDERATIONS

Another key factor is product reliability. Products that break often or that are down for a long time earn their manufacturers a bad reputation in an extremely short period of time. With reliability engineering people on the design team, reliability can be designed in as an integral part of the product's performance requirements.

Some questions to keep in mind during product design include the following:

- Is the mean time between failures (MTBF) adequate?
- Is the mean time to repair (MTTR) acceptable?
- Which is more important?
    - Absolute reliability
    - Actual availability
- What is the repair philosophy?

Implementing some concurrent engineering technical guidelines can sometimes have a very minor negative impact on product reliability. On the other hand, they can significantly improve product availability. Which is more important to the customer for this particular product? The repair philosophy for the product (unit replacement, board swap, throwaway) will also have an impact on the reliability requirements and the packaging and layout of the product.

COMPONENT SELECTION CONSIDERATIONS

Component selection (or design technology selection) considerations are also very important. They have a big impact on both the up-front design engineering times and costs and on the ongoing production and test costs of the product. In the next few paragraphs, we'll look at the relative advantages and disadvantages of each of the component technologies—full custom, standard cells, gate arrays, programmable logic, and standard logic chips—available for electronic designs. Keep in mind that one of the tenets of concurrent engineering is to use the simplest technology consistent with product performance needs.

Full Custom Designs

The following list outlines the salient features of designs that utilize fully custom ICs:

- Very high design costs
- Very long design time (months)
- Redesign prohibitive
- Long test development
- Low prices at high volumes
- Maximum number of features per design

Because of the very high "iteration cost" for a full custom IC, great care must be taken during design to insure that the chip works right the first time. Thus, extra time spent for design verification, logic simulation, test vector generation, fault simulation, and race and hazard (timing) analysis is really "insurance" against the even longer amount of time that would be required in the event of a redesign.

Full custom designs are most appropriate where products will be produced in high quantities or where there is no other way. (Question: Is there really no other way to fit the required number of performance features into the design?)

Standard Cell Designs

ASICs made from standard cell technology share some of the characteristics of full custom designs. They are usually inexpensive in high quantity situations and can include a great number of performance features. They are not quite as expensive or time consuming to design as a full custom IC, but it may still take several weeks and a significant amount of nonrecurring engineering cost. Standard cell characteristics include the following:

- High design costs
- Moderate design time (weeks)
- Redesign very expensive
- Long test development
- Low prices at high volumes
- Many features per design

Redesign is again a very expensive and time-consuming proposition, so getting it right the first time is critical. Test generation times can also be very long with standard cell designs, although some standard cell vendors are including on-chip testability features that allow them to reuse test vectors for the major cells in the application-specific IC. Those features do not, however, eliminate the need for custom test vectors for the user-specific portions of the IC, and they are not of much help in board level testing (once the IC has been integrated into its "next assembly").

Gate Array Designs

Gate arrays often represent a good compromise between using standard (or programmable) logic devices and going the full custom or standard cell route. Gate array designs can be characterized as follows:

- Moderate design costs
- Moderate design time (weeks)
- Redesign expensive
- Moderate test development
- Medium prices at high volumes
- Many features per design

Gate arrays are not as inexpensive as full custom or standard cell designs in very high quantities, but for medium volume situations that may not be a significant factor (when comparing the total of NRE and recurring component costs for the total quantity of components to be purchased over the life of the product).

Design time is relatively short for a gate array, but an iteration is still expensive and time consuming. Gate arrays offer many features per design, and their test development time and cost is typically lower than that for a full custom or standard cell design (particularly if the gate array makes use of one of the on-chip testability techniques discussed in a later chapter of this text).

Programmable Device Designs

Programmable devices, such as programmable array logic (PALs), electrically programmable logic devices (EPLDs), and logic cell arrays (LCAs), offer another set of design for performance options. They have the following characteristics:

- Low design cost
- Short design time (days)
- Fast and inexpensive redesign
- Inexpensive device test development
- Medium prices at high volumes
- Moderate number of features per design

Since programmable devices are user-configurable, both the initial design time and any iteration time are considerably shorter than can be achieved using any of the custom device options. Depending upon the device type, test programs may already exist (for the unprogrammed device) or may be relatively inexpensively generated.

Programmable devices don't hold near as many features as gate arrays or full custom devices, but do allow a lot of "glue logic" to be incorporated in a design without having to utilize many discrete standard logic device packages. Designers using programmable logic devices should pay particular attention to the design for testability guidelines that will be covered later in this text.

Standard Logic Device Designs

Standard logic devices, of course, will not go away completely anytime in the near future. These building blocks are inexpensive to use in designs (except in very high quantities, where an ASIC might be more appropriate) and readily available. The major characteristics of these devices include the following:

- Low design cost
- Short design time
- No device redesign
- Already existing test programs
- High overall costs at high volumes
- Minimum number of features per design

Prototype designs can be modified quickly and easily (although PCB layout iterations can be costly and time consuming). Test programs for the devices themselves already exist in many device and board tester libraries. The main disadvantage, of course, is the minimum number of features contained in each device.

Most electronic printed circuit boards will use a mix of custom, programmable, and standard logic devices. Care should be taken to select testable versions of the devices whenever they are available and to follow the design for manufacturability and design for testability guidelines for printed circuit boards contained in later chapters of this text.

PERFORMANCE ENHANCEMENT TRENDS

There is no end in sight to our ability to cram more functionality in less space. This is OK, as long as we remember that we have to be able to produce products, verify their performance to extremely high-quality levels, and fix them when they break.

The chart in Figure 5-2 shows the forecasts for semiconductor device density over the next decade, while Figure 5-3 illustrates the forecast for interconnection pin pitches on packaged (and sometimes unpackaged) semiconductor devices. We will clearly have to employ all of the resources available to us in order to cope with this increased complexity.

One of the things we want to make sure we do when making design for performance and design (or component) technology decisions is to use the lowest level of technology possible—consistent with competitive requirements such as performance, perception of the product as "high tech," and so forth, because both

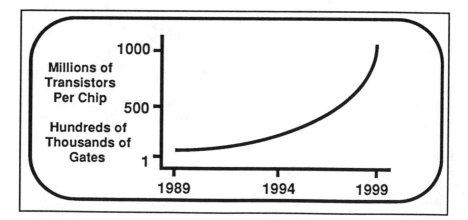

FIGURE 5-2   Gate count trends

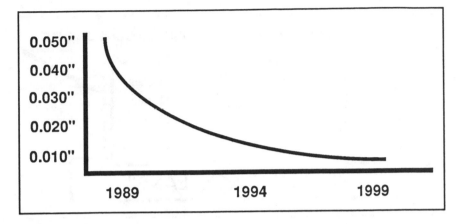

FIGURE 5-3   Pin pitch trends

assembly and test techniques will be challenged as components begin appearing with ever-finer spacings between pins.

Packaging technologies are also undergoing fundamental changes. The work done in the 1980s by Trilogy, Mosaic Systems, and Wafer Scale Integration is beginning to pay off in the form of products available from the technologies they pioneered.

Illustrated in Figure 5-4 is the multi-chip module, or 3-D packaging, concept. Design one of these without consulting purchasing, manufacturing, quality, test, and service. What will be the result? Most likely a very clever design that is so complex it can literally never be built economically in volume production or tested adequately (or at all!).

The only solutions are to design products and processes in parallel. If we outstrip the manufacturing process, our new designs may never see the light of the marketplace. The same is true if we outstrip our test capabilities. Equal engineering efforts must be applied to improving manufacturing and test technology as they are being applied to new product technology. At the same time, products must be designed to be easier to build and easier to test. This means concurrent engineering.

SUMMARY

Design for performance is certainly one of the most critical, if not the most critical, element of product design. And while product performance must be achieved within the size, weight, power consumption, and material cost goal constraints of the organization, design for manufacturability, testability, and serviceability can-

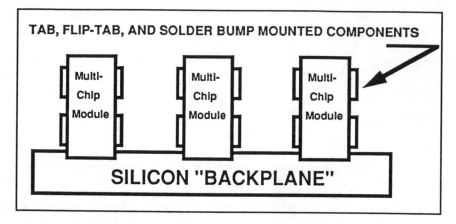

FIGURE 5-4   A multi-chip module

not be ignored. In the old serial engineering environment, performance and functionality were typically the only concerns of the product designer. Manufacturing, test, and support were "someone else's worry."

In the concurrent engineering environment, required for success in the future, that attitude will no longer suffice. Product design engineers will have to make quantitative trade-offs and avoid technology decisions based on opinions if they are to ensure that they can meet not only performance goals but also reliability goals, quality goals, manufacturability goals, testability goals, and product compliance requirements. This will take teamwork by a product development team that has all the right expertise and that can apply it at the right time—during product design—so that the final design is right the first time.

REVIEW

1. Product speed decisions should be based on (a) *the fastest possible speeds that can be achieved* or (b) *the fastest speeds required by the customer application.*
2. Product size for new product designs that will not be "retrofitted" into existing spaces are (a) *flexible* or (b) *fixed.*
3. Product quality, reliability, and compliance considerations should be taken into account (a) *once the product design is fairly well firmed up* or (b) *while the conceptual, specification, and initial design is being done.*
4. The choice of which component technologies to use in a new design should be based on (a) *nonrecurring costs alone,* (b) *recurring costs alone,* or (c) *the combination of (a) and (b).*

5. Leading edge packaging technology should be used (a) *when it is necessary to gain a competitive edge in the marketplace* or (b) *because it is leading edge.*

Answers: 1: b, 2: a, 3: b, 4: c, 5: a

NOTES:

6

Computer-Aided Engineering and Test Considerations

In the previous chapter, we talked about the performance issues of our new designs and the building blocks we use to implement them. In this chapter, we'll discuss the tools that can be used during the design, design verification, and product verification (i.e., production testing) phases of the concurrent engineering process. The major topics that we are about to explore include the following:

- The current design process
- The concurrent design process
- CAE tools available for concurrent engineering
- CAE tools needed for concurrent engineering
- Design verification trade-offs
- ATE test techniques
- Trends in components and CAE tools

THE CURRENT DESIGN PROCESS

The current design process typically utilizes CAD/CAE tools as "islands of automation" in most organizations. There are some tools available for schematic entry, some for logic simulation, some for fault simulation, some for PCB layout, and so forth. Most tools today are aimed at the ASIC design marketplace, with work only now commencing in earnest on the previously neglected board and system levels of design.

There are some links available to transmit design and test data from one tool to another, but these are sometimes cumbersome and they are by no means complete for the board and system level designer. Thus, it is difficult to "uplink" design data from one level of integration to another and to "download" design data to the manufacturing operation. And the tools don't really support the "-ilities" during the design process. Rule checking is usually an after-the-fact operation, if it is done at all (because of "time to market" pressures).

Design engineers, working in their own little world, sometimes on different floors of the same building and sometimes in different buildings (or cities, states, or countries), create new product designs without worrying much about how the product will actually be built, tested, and serviced, or the impact that some of their design decisions will have on the total quality of the final product. This situation must change if companies are to maintain (or regain) a competitive stance through the efficient execution of the concurrent engineering process. "Frameworks" where multiple tools from multiple vendors can all operate "seamlessly" at multiple levels of design integration are being talked about and are currently being developed and refined.

THE CONCURRENT DESIGN PROCESS

Contrast the previous description with the one illustrated in Figure 6-1. This drawing shows the concurrent design engineering process concept. This model applies to the concurrent design methodology, whether it is implemented manually or with CAE tools. Supporting the design activities are the manufacturing, test, quality, and service engineering functions. The support functions "run in the background," but concurrently and in real time with the design activities.

Concurrent engineering is not capital intensive. It is communication intensive.

That communication intensiveness is characteristic not only of the interplay

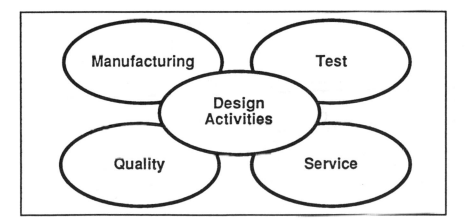

FIGURE 6-1   The concurrent engineering model

between the actual design activity and all of the support activities, but also of the hierarchy of design activities.

The drawing in Figure 6-2 illustrates the concept of constant communication between designers at different levels of integration. System requirements are allocated to subsystems and boards. Board level requirements are allocated to the component designers. Design data, including physical descriptions and test information, flows back and forth as the give-and-take of the concurrent engineering trade-off process continues and the final product design evolves.

In an ideal world, all of the workstations and tools would be linked, as shown in Figure 6-2. As mentioned earlier, work is proceeding apace on accomplishing the integration of design tools into frameworks that will allow for the automation of the concurrent engineering process. Until that work is completed, however, it will take improved human communication. And that can be accomplished regardless of organizational size or structure, given the right leadership and commitment from top management and the education and cooperation of each of the members of the concurrent engineering product development teams.

CAE TOOLS AVAILABLE FOR CONCURRENT ENGINEERING

If we look at the tools available for concurrent engineering today, we see a list like this:

- ASIC level design and simulation
- Board level design and simulation

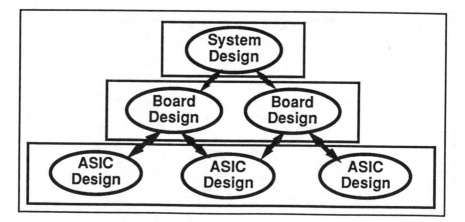

FIGURE 6-2   CAE tool link hierarchy

- PCB design CAD (computer-aided design)
- ATE systems
- "Linking" software from *some* design platforms to *some* test platforms at *some* levels of design integration

There are specific tools for each specific level of design integration. There are specific tools to perform specific tasks at each level of integration, and some of them run on common platforms.

There are also some (usually one-way) links between the hierarchical levels and some links between the tasks. What's needed, in terms of CAE support of concurrent engineering, however, is *not links*. It is *integration!*

CAE TOOLS NEEDED FOR CONCURRENT ENGINEERING

Some of the design automation items necessary for the automated, rather than manual, implementation of the concurrent engineering discipline are listed in Figure 6-3. It is encouraging to see articles in the technical and business press that development of at least some of the items on the list is beginning.

Professional organizations like the IEEE are also getting involved in trying to standardize many of the items listed in Figure 6-3 on frameworks that support open architectures. But until these tools are actually developed, debugged, and deployed in most organizations, organizations practicing concurrent engineering will have to work with the tools available while encouraging their design automation suppliers to develop the items they need for truly automated CE.

Tools Needed for Successful Automated CE

- **CAE Frameworks That Support ALL Engineering Tasks**
- **Library Elements Than Can Be Reused At Various Levels of Integration**
- **Common Languages and Formats for Design Information**
- **Common Data Bases and File Servers at All Integration Levels**
- **Synthesis Tools That Include CE Expert System Rule Checking**

FIGURE 6-3   CAE tool requirements for the future

In addition to the elements listed in Figure 6-3, which really describe what are traditionally thought of as tools for the "design-oriented" tasks, are the other tools that include the concurrent engineering elements within the framework. Some of these additional requirements include:

- "Seamless" inter-tool communication
- Support for multiple tools
  - Design, verification
  - Layout, packaging, and thermal
  - Manufacturing, test, and service
- Data management
- Design representation
- Methodology management
- "Transparent" user interfaces

The list above covers the actual design task (via inter-tool communication for schematic capture and logic simulation), the support tools (i.e., thermal, layout, packaging, manufacturability, testability, and serviceability rule checkers), and documentation and data management. Many of these pieces already exist in proprietary forms. Some also exist in the commercial marketplace, but are not integrated together in such a way as to form a "desktop" where all of the tools are "transparent" to the actual concurrent design engineering process.

Another real need is for a set of standard library elements that can be used at multiple levels of integration. Elements such as transistors and gates can currently be moved into the component realm, but moving component data into the circuit (board) environment is not smooth. And almost nothing exists to bring component and circuit data up to a system CAE level, and vice versa.

A critical element in making this integrated environment meet the engineering needs that have been mentioned is the ability to describe systems at a very high level and still be able to use and understand the synthesized structures at lower levels. It's also critical that optimization changes made at lower levels be able to be back-annotated up to higher levels so that descriptions at all levels remain accurate and interconnected.

Common languages and data bases are also needed. So far, EDIF, the Electronic Design Interface Format, and VHDL, the VHSIC Hardware description language, have been fairly fully developed and are being supported by many CAE and ATE vendors. The focus, however, has still been at the device level. A common language and common data base, one that again provides the ability to transcend levels of integration and CAE workstation platforms and operating systems, would go a long way toward allowing the full automation of concurrent engineering.

Synthesis tool development has already begun at the device level. Similar tools will eventually be developed at the board and system levels, and they will hopefully use the same language and data base as the chip level programs. All of the abstraction levels—architectural, behavioral, register, gate, and physical—must be supported, and expert systems must be developed to automatically check for (or synthesize in) the "-ilities." These systems should allow customization by the user so that the expert system knowledge base can be constantly improved, as well as both forward and reverse annotation capabilities.

DESIGN VERIFICATION TRADE-OFFS

Design verification trade-offs also need to be considered with respect to the concurrent engineering process. Hardware design verification, using a prototype of the actual design, will of course continue to be used, but in the concurrent engineering process we wish to simulate the design to make sure that the hardware will work correctly the first time.

Systems exist to provide complete design verification in software at some levels of integration. Others make use of programmable hardware to perform the design verification without the need for an actual device or board hardware prototype. Getting models for new, complex devices, however, still remains a problem. Thus, what happens many times is that a combination software/hardware design verification approach is taken.

Some of the advantages and disadvantages of hardware design verification are the following:

- It requires design verification (TEST) vectors to exercise the design.
- It requires pin electronics to physically apply and receive test vectors.
- It requires that physical prototypes be fabricated.
- It allows real world verification.
- It helps proof fabrication processes.
- It may require hardware design iterations.

One of the major disadvantages of hardware design verification is the need to construct an actual prototype in order to perform it. Another major task is the development of a comprehensive set of design verification patterns (which are actually test vectors).

On the good side, having prototype hardware helps in proofing processes and allows for "real world" verification of design performance. Pin electronics of some type, either as an adjunct to the CAE system or in the form of an ATE system, are required and, if the design does not work right the first time, design iterations may be required.

Software design verification also requires the development of the stimulus vectors that will be used to test the functionality and performance of the design. Many of the design for testability guidelines presented in Chapter 8 can help ease the task of developing the test vectors considerably. Some of the advantages and disadvantages of software design verification are the following:

- It also requires design verification (TEST) vectors.
- It requires accurate models.
- It requires good CAE tools.
- It provides no help in proofing fabrication processes.
- It makes iterations fairly quick and easy.

The main drawback to using software design verification, at least above the device level, is the lack of availability of accurate device models. Using software makes iterations fast and easy, but is of no help in proofing processes. Finally, good CAE tools (like those described previously as "needed") are required in order to do software design verification quickly, efficiently, and accurately.

Combination hardware/software design verification is probably the most widely used method (above the chip level) today. It is basically a compromise in order to overcome the lack of availability (or difficulty in creating) accurate models of very high complexity devices or of merchant devices (such as microprocessors) for which models are not released by the supplier.

The requirements and characteristics associated with combination hardware/software design verification areas follows:

- It requires design verification (TEST) vectors.
- It requires the equivalent of pin electronics.
- It uses design verification hardware to emulate real prototype hardware.
- It requires accurate libraries.
- It requires good CAE tools.
- It makes iterations easy.
- It is of little help in proofing processes.

Regardless of the method used to verify the functionality of the design (e.g., the correctness of its operation), care should be taken to make sure that any features "temporarily added" to the design to ease the verification task become permanent features. This will be especially important for production testing, as anything that eases the design verification task also eases the tasks of high fault coverage test program generation and troubleshooting, using commercial automatic test equipment (ATE), or diagnostics, using built-in test techniques.

It is also important to be aware of the limitations and costs associated with the production assembly environment into which the product design will go. Automatic insertion equipment, for example, may have maximum possible board (or panel) size limitations that, if they are exceeded, will not allow a design to utilize automatic insertion equipment. Hand stuffing is then the only alternative and is much more expensive than automatic parts installation.

Pick and place equipment may have component package type limitations or package size limitations. If these limitations are ignored, secondary (manual) operations will have to be performed after automated parts placement. It can certainly be done. The question is: At how much cost and in how much time? Care should be taken to understand how the product will be assembled when it is designed.

ATE TEST TECHNIQUES

Once design verification has been completed and a design is released to manufacturing for production, each printed circuit board must be tested in some manner. The four most commonly used commercially available types of board testers used to perform the testing tasks in a typical production environment include:

- Manufacturing defects analyzers
- In-circuit testers
- Functional/performance testers
- Combinational testers

Some organizations use only one of the above types of testers. Others may use more than one, and still others may use none of them, preferring to build test equipment in-house or use "hot mock-ups" of the final product into which a board will be inserted as a test bed.

Since, however, these testers are the types most commonly used, and since at least cursory knowledge of their philosophy, methods, and technology is needed in order to understand the importance of the testability guidelines contribution to concurrent engineering, we will explore each one briefly.

The **manufacturing defects analyzer** (MDA) is also sometimes called a manufacturing defects tester (MDT). As its name implies, it is designed to find defects that occur in the product due to the manufacturing process. One might ask, of course: Why are you manufacturing defects (instead of good products!)? This is a very good question. If you follow the concurrent engineering guidelines, and if your factory makes use of statistical process control information to reduce the number of manufacturing defects to a very low level, the need for this type of tester in the production flow can be eliminated.

The philosophy behind the development of the MDA is as follows: Most defects in devices, and especially in printed circuit boards (or other similar assemblies) and even final systems, are introduced during the manufacturing process itself. That situation is, unfortunately, still all too true in most factories. The MDA was developed because in-circuit testers (to be discussed next) grew as complex and expensive as the functional testers they were designed to take the place of. An MDA is an inexpensive piece of equipment to purchase and program, although fixturing, especially for surface mount and fine pitch technology (SMT and FPT) board designs, is becoming very expensive in most cases and problematical in others.

The best use of an MDA is in the early production phases of a product when the number of manufacturing defects is higher than desired. The MDA provides the information necessary to clean up the manufacturing process and eliminate the MDA testing step for that board (or family of boards). The MDA uses a bed-of-nails fixture (see Figure 6-4) to contact each (accessible) node on the board. It looks for connections that should be present but are not (opens), connections that exist but should not (shorts), and missing, wrong, or wrongly inserted components.

Some MDAs can also check resistor and capacitor values. Most do not check the digital ICs on the board at all (except sometimes for IC pin impedances to determine if a digital part appears to be present and appears not to be inserted wrongly).

MDAs use stimulus voltages on the order of 100 millivolts in order not to turn on transistor and IC junctions that would look like short circuits. Testing is done without normal operating power applied to the circuit under test. Some testers use strictly DC stimulus and measurement techniques. Others may use AC techniques for nodal impedance measurements. In either case, the main objective is to find the gross manufacturing defects—opens, shorts, and missing and wrong components.

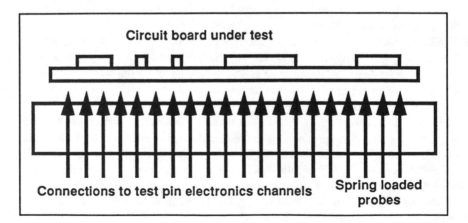

FIGURE 6-4   A bed-of-nails test fixture

The benefits the MDA provide are: isolating process-induced shorts and opens at an early stage in the manufacturing process, requiring a small programming effort, and requiring minimal test operator skill levels. The limitations include: little or no parametric testing capability and no functional test capability.

The **in-circuit tester** was developed in the late 1970s as an alternative to the functional board tester. Functional board test programming was getting expensive as design complexity increased and test engineers failed to get designers to implement basic functional design for testability guidelines. Finding faults with a functional tester is also a slow process, and lots of faults were built in those days. We still build far too many, which is why the in-circuit tester is still so popular.

The in-circuit tester is a good process monitoring tool, as was the MDA. But the objective again is to get first pass board yield high enough to eliminate the need for the in-circuit tester (except, perhaps, as a diagnostic tool for defective boards screened out at the functional test level). The philosophy of the in-circuit tester is: If all of the components are the correct components, all are of the correct value, and all are correctly installed and connected, then the circuit assembly will function correctly. And this assumption is mostly true.

There are defects, particularly functional interactive and timing defects, however, that the in-circuit tester will not find. Thus, the possible fault coverage with an in-circuit tester usually lies somewhere between 85 percent and 95 percent (depending upon the unit under test's testability features, the level of effort spent writing the test program, and the ability to fixture the board).

The in-circuit tester was also developed because most board level defects are introduced during the manufacturing process itself, and this is still all too true. We must change that situation, and concurrent engineering's focus on fitting product design to production processes can help in this regard. In-circuit testers can provide excellent information about the causes of defects, and advantage should be taken of that fact to solve the manufacturing problems once and for all.

The in-circuit tester also uses a bed-of-nails fixture for access to as many nodes on the board as can be contacted. An opens and shorts test is typically the first test performed using an in-circuit tester. Then the value of each component is measured. Finally, a power-on (usually limited) functional test is applied to each individual component. Because it can functionally exercise each component (or group of components, where individual components cannot be isolated for in-circuit testing due to lack of testability), the in-circuit test (again depending upon the level of programming effort) can find most of the defects that can occur on a board.

The ICT typically uses the same low-level stimulus and measurement technique as the MDA for its opens and shorts testing phase. It then employs sophisticated analog guarding and high-speed, low–duty cycle digital back driving techniques to functionally test the components on the printed circuit board.

The benefits of the in-circuit tester include: One-pass diagnostics for process-induced faults, relatively easy programming (as compared to functional or combination testers), and moderate operator skill level requirements. The limitations include: inability to check some component tolerances, inability to test some components at all (due to circuit design factors), inability to test most component interactions, and limited functional test capability.

Functional testers were the mainstay of factory and field testing from the late 1960s through the middle 1970s. They are reemerging, under the new name "performance testers," in the early 1990s. People have discovered that when the manufacturing process runs correctly, with first pass board test yields in the 90 to 95 percent-plus ranges, functional board testing is less expensive than in-circuit testing. After all, why spend an extra minute or two in handling and test time on the in-circuit tester for every board when only 5 percent or so of the boards have defects? The functional (or performance) test is done first—only boards that fail go to the ICT (or MDA) for fault diagnosis.

While performance testers may be more expensive than in-circuit testers (although they do not have to be), they execute tests more quickly and provide for better fault coverage. The functional (or performance) tester tests a printed circuit board assembly as an entity, rather than as a collection of individual parts, on the theory that, if the assembly functions correctly, then all of the components must be the correct components of the correct types and values and they must all be installed and connected correctly.

That assumption is also "mostly true." There are some faults (e.g., wrong value pull-up resistors, out of tolerance components that do not affect performance that the performance tester will miss (and the in-circuit or combinational tester (to be discussed next) can find).

Fault coverage on a functional tester that runs at full unit under test speed (or at least close to it) can easily be in the 98 percent-plus range (given that the board configuration allows that and that a comprehensive test program has been developed). Programming, however, can be expensive (again depending upon unit under test testability considerations). While functional testing may be more expensive than other methods on its own, the higher yields at the system level normally pay for the difference many times over.

Due to the high circuit speeds of today's modern boards, edge connector access (rather than bed-of-nails fixture access) is preferred for functional (or performance) testing. This keeps wire lengths between the tester and the circuit under test as short as possible. Stimulus vectors may be applied to flush out all "stuck at 0" and "stuck at 1" faults (as determined by an automatic test pattern generation program and/or interactive test vector generation using a fault simulator). Alternatively, system emulation testing may be applied to exercise the functionality of the board—preferably at high speed to catch all of the functional interactive and timing faults. Most performance testers include hardware and software tools to guide a test operator through a probing sequence in the event that a test fails in order to isolate the fault.

Today's functional testers incorporate the latest in SMT and FPT design techniques and run at very high speeds. Testers with 40 MHz data rates (80MHz in the multiplexed mode) are available to provide true performance testing capabilities. There is local memory behind each pin (up to 64K bits and sometimes more) to handle the large number of test vectors usually associated with complex board designs.

These testers can synchronize to multiple clocks, have support channels for device and board level scan design techniques, and, because they can execute millions of patterns at high speed, can detect most "soft failures." Those are the benefits. The limitations include: high initial purchase price, long program generation times (unless units under test are optimally testable), and the requirement for fairly skilled test engineering resources.

The final ATE category to be discussed in this chapter is the **combinational tester** category. Combinational testers are testers that contain the capabilities of both a sophisticated in-circuit tester and a performance (i.e., high speed functional) tester. The theory is that, even though the machine is more expensive to purchase and to program, boards will only have to be handled once, fault coverage will be as high as possible, and testability problems can be overcome.

Those statements are mostly true in today's environment. But once concurrent engineering is a widespread practice, lower cost high-speed functional testers, augmented by sophisticated expert systems for diagnosing the few remaining faults on boards, will become the norm. There are already low cost alternatives available for testing boards that incorporate testability bus architectures.

Most combinational testers provide the option of which order in which to run the tests. In a typical production operation using a combinational tester, boards with high yields (i.e., very few defects) would undergo the functional portion of the combinational test first. The good boards go to the next assembly stage while the in-circuit test and diagnosis is invoked, in order to locate the defect causing the fault on the faulty boards. For low-yield boards, the in-circuit portion of the test is done first and a final functional test is performed once the majority of the defects have been removed from the board.

Combinational board testers usually use bed-of-nails fixtures. Some use the edge connector(s) of a printed circuit board plus a limited bed-of-nails fixture. Others use what is called "dual stage fixturing," where only some spring pins contact the board for high-speed testing and the rest of the pins are applied for fault isolation, using in-circuit testing (or for a preliminary opens and shorts test, after which they are disconnected from the board under test to allow for performance testing).

Most combinational testers have high pin electronics counts—on the order of

2,048 in-circuit pins (or more) and sometimes as many high speed pins. That's one reason they are so expensive. New scan support channels and the software to support them are being introduced and will continue to appear as we move into the 1990s (and as more testability bus standards reach the approval and implementation stages).

When unit under test speeds exceed tester speeds, two tester pin channels, interleaved (i.e., multiplexed), can sometimes be used to extend the tester's capabilities. When node counts exceed tester capabilities, fixtures can be equipped with testability circuits to expand the number of nodes that can be contacted by a finite number of tester pins.

TRENDS IN CAE AND ATE TOOLS

There are significant trends in the electronics industry that both make the concurrent engineering discipline more important than ever as the complexity of products continues to increase—sometimes on what seems an exponential basis—and bode well for the future of automating the concurrent engineering process. There will be more open architecture frameworks and more sharing of libraries, even if they must be developed by third parties and the individual models paid for. Our ASIC foundry tools sets will be better able to communicate better with other tools at higher levels of integration, and we'll see concurrent access to design data bases in real time.

Much of the "dog-work," both in design and in the implementation of the "-ilities," will be automated with design compilation and expert systems. Those are exciting prospects as long as we use concurrent engineering now to make sure our companies are still in business to take advantage of them! The platforms to support those exciting developments will continue to become more powerful and more affordable. They will thus be used by the "-ility" engineers to support design activities in real time.

The ATE business as we know it today will change significantly as tester functions migrate to workstation environments and built-in tests begins to replace huge, expensive testers designed to "overcome" today's testability problems.

Even process control will migrate to the workstation environment, in the fully networked "business of the future," to allow immediate feedback to design activities on what works in the factory and what doesn't and to manufacturing cells on what they're doing right and wrong.

As we move into the 1990s, the following trends are becoming ever clearer:

- More competition on a global basis
- Increased attention to packaging technologies
- Increased attention to built-in self-test

- Increased attention to process control
- Increased attention to improved product quality
- Increased focus on total customer satisfaction
- Increased implementation of concurrent engineering

Companies that do not recognize these trends and take immediate action to begin the process of coping with them, hopefully on a proactive basis, may not be around a few years from now. Let's hope we all work for the forward-looking, long-term committed organizations that will be the leaders in the next decade.

SUMMARY

Current CAE tools are typically used as islands of automation to speed various design engineering tasks. In a concurrent engineering environment, these islands need to be connected together to facilitate improved communication—either electronic or human or both. Current developments will facilitate automating many of the concurrent engineering technical tasks (i.e., design for manufacturability, testability, and serviceability, among others) that currently require personal engineering attention.

New tools will help in doing design verification tasks that will result in "right-the-first-time" designs and drastically reduce the number of design iterations required in a typical serial engineering environment. These new tools will also help to ensure that new product designs fit into company manufacturing and test processes, either existing or evolved to support new technologies. As new design innovations continue to occur, the cost of acquiring and using tools will continue to come down and the levels of integration will increase. Total quality management concepts will be applied to continuously improving the product design process and the integration of products and processes.

REVIEW

1. Truly automated concurrent engineering requires (a) *links between design and process systems* or (b) *integration of design and process systems.*
2. Concurrent engineering is (a) *capital intensive* or (b) *communication intensive.*
3. The most common requirement between hardware, software, or combination hardware/software design verification is (a) *circuit models,* (b) *hardware prototypes,* or (c) *test vectors.*
4. Automatic test equipment should be used to (a) *efficiently locate defects* or (b) *identify the sources of defects, both manufacturing and design, so that they can be prevented in the future.*

5. Concurrent engineering success is achieved by taking a (a) *proactive stance* or (b) *a reactive stance.*

Answers: 1: b, 2: b, 3: c, 4: b, 5: a

NOTES:

7

Design for Manufacturability Considerations

Thus far in this text we have been concerned with the people, organizational, and tool issues of the concurrent engineering process. In this chapter (and the next two), we are going to discuss a large number of technical "design for" guidelines that, if followed rigorously whenever possible, will help make a design easier to manufacture (and easier to manufacture correctly the first time), easier to test, and easier to service.

The design for manufacturability guidelines don't normally impact product performance requirements very much. But they do need to be considered early in the design phase of a product and their implementation requires close coordination between circuit design, printed circuit board design and layout, product packaging, and product manufacturing engineering functions.

PRODUCT PARTS COUNT CONSIDERATIONS

Product parts count considerations are just as important to the manufacturing organization as they are to the design engineering organization. The first guideline is: *keep the total number of parts to a minimum*. Sometimes, certain design for manufacturability guidelines conflict with the design for testability and design for service guidelines to be presented in the following two chapters. The key is to be aware of the interrelationship between the guidelines and to make design engineering decisions that will integrate the best possible compromise in each case.

Keeping the *number of part types to a minimum* is also important, as is *standardizing on reusable modules*. Following these guidelines can lead to reduced development times, fewer purchase orders, lower inventory levels, and faster order cycle times.

Many organizations have standard (or at least preferred) parts lists that designers must use for new product designs. This can sometimes be a little restrictive, but the benefits can be very large. Using the same part in multiple designs lowers component purchase costs, and the higher volumes mean more leverage with the part supplier in terms of quality levels. Parts used for long periods of time are also better understood. Thus, there is less likelihood of discovering a component idiosyncrasy late in the product development cycle.

Using standard packages is also important if we are to fit the product into the process. The fewer the requirements for new tooling or manual operations due to the lack of tooling for automated processes, the lower the manufacturing cost and the less likelihood for errors.

DESIGN FOR ASSEMBLY

Design for assembly (DFA) guidelines have come a long way, as many firms have recognized the savings (like those pointed out in the introduction to this text) that can be achieved through the concurrent engineering process.

Listed below are six simple guidelines that can drastically reduce the time it takes to perform assembly operations on the manufacturing floor. There is, in fact, great room to exercise your creativity in coming up with unique ways to make parts do double duty, for example, or in eliminating fasteners through clever packaging design.

- Reduce the number of screws required
- Reduce the number of screw types required
- Eliminate other fasteners where possible
- Minimize settings and adjustments
- Use "snap-on" or "snap-together" techniques
- Make parts serve multiple purposes

In addition to following these guidelines, one of the most important factors to take into account is the sequence of assembly operations. Shown in Figure 7-1 is an example of an assembly that is put together once, with access from only one direction required. This type of design facilitates automation. Presuming that each item added is indeed the right size and shape, and that it is functioning properly (i.e., it has been tested before being included in the next higher level of integration), the completed assembly should function the first time, every time (or at least a very high percentage of the time).

On the opposite end of the spectrum is a design that requires already installed

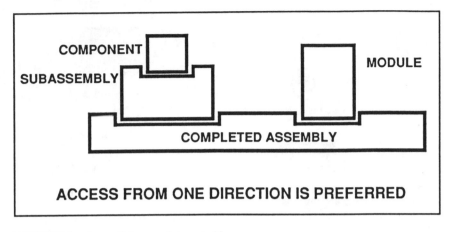

FIGURE 7-1   Sequential approach to assembly

items to be removed in order to install a new item (and then to reassemble what was already assembled before it was disassembled). Remember that every operation removed from the process reduces the chance for mistakes and lowers costs.

Since a good portion of the labor cost in most factories is spent on rework, preventing errors is a high leverage activity. Some of the ways that design for assembly guidelines can be used to facilitate error-free assembly of mechanical product components are to:

- Make parts self-aligning.
- Design for symmetry where possible.
- Increase asymmetry where appropriate.
- Add external features to prevent improper assembly.

Examples of these techniques are illustrated in Figure 7-2.

As shown in example Figure 7-2(a), one way to reduce the possibility of assembly errors is to add features to parts that will make them self-aligning. This not only helps in making sure that the parts go together right the first time, it also helps to prevent possible damage at the next assembly level due to wrongly installed parts when an assembly is tested. The illustration shows how a mating pin and hole for the parts on the right have been made a different size than the other three. The parts will thus be easier to assemble, and any chance of the key between the two parts being in any of three of the possible wrong positions will be eliminated.

Another technique that can often be used is to design parts, especially where multiple parts are used in the same assembly, for symmetry. This reduces the

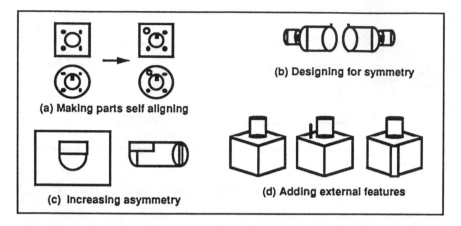

FIGURE 7-2   Some error-free design for assembly techniques

number of individual part designs required. The example in Figure 7-2(b) shows symmetrical pins that can be used on either the left or the right side of an assembly. In other cases it may be more desirable to increase the asymmetry of a component to prevent improper installation. Where parts must be specifically placed and should not be allowed to be interchanged (even if the assembler tries to use excessive force in an attempt to do so), increasing asymmetry as illustrated in Figure 7-2(c) will prevent it. External features may also be added to mechanical components to make it easier to assembly them correctly (or more difficult to assemble them incorrectly, depending upon your point of view!).

In the example of Figure 7-2(d), a guide pin has been added to the top of the flat surface of the cube so that the component coming down on top of it is easier to align and install correctly. In the example on the right, one corner of the cube has been flattened so that it will only fit into the assembly below it in one direction. Combining both techniques is also not a bad idea, but keep in mind that we don't want to unnecessarily complicate the individual component tooling requirements or fabrication processes.

ELECTROMECHANICAL CONSIDERATIONS

Electromechanical issues can also be important to the factory and the field, especially when one considers the increasing globalization of the electronics marketplace. There are even issues of material specifications for connections between components and subassemblies. Low insertion force connections, for example, normally require noble metals (i.e., gold) that will not oxidize and that will have very low electrical resistance. Higher insertion force items also require careful materials selection to insure gas tight connections and to prevent the occurrence of corrosion on a long-term basis.

With all of the different primary (mains) power sources around the world, and with modern power supply design methods available to almost everyone, it is amazing how many products are still designed with the requirement that jumpers be configured (either in the factory or, worse, in the field) in order to accommodate differing input voltages. Many products now come with user-settable switches, which is better than having to remove covers to change jumpers. But it still leaves lots of room for error (and the potential destruction of the product). The preferred method is to *design power supplies so that they can automatically adapt to differing input voltages.*

A similar situation exists in the area of power cords. Plugs and sockets differ widely at the main power input end. They do not, however, have to vary at the product end (if the product power supply was designed in accordance with the previous guideline), and power cords are thought of as modular design components. With the modular power cord approach, one product design and a set of (inexpensive) power cord types can accommodate a worldwide customer base. The alternative is to design, assemble, and inventory multiple complete product configurations. This is expensive and it reduces your ability to respond rapidly to changing customer demands.

Another design for assembly guideline is: *minimize the number of cables internal to the product.* Cables are expensive to design, fabricate, test, and install. They also tend to cause reliability problems. Thus, minimizing the number of internal cables in a design can save money in production and result in fewer field failures.

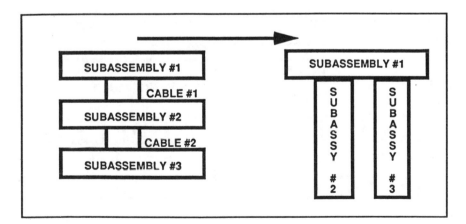

FIGURE 7-3   Minimizing the number of internal cables

In the example of Figure 7-3, the product has been redesigned to eliminate the cable altogether. Subassemblies number 2 and 3 now plug directly (and from the same direction!) into subassembly number 1. Circuit performance may also be enhanced due to shorter lead lengths and reduced contact resistance (i.e., the number of connections subject to resistance has been halved). Costs have certainly been reduced.

PHYSICAL MODULARITY

There are several categories of (primarily) mechanical design for manufacturability guidelines that it is also important to be aware of during the product design phase. Some of them, such as product size and weight, are usually customer- (or marketing-) driven external considerations. Others, such as accessibility, connector implementation, and printed circuit board layout, are features that are more important to the design engineering organization's "internal customers" (i.e., the manufacturing, test, and service organizations). One of the most important considerations is physical modularity (see Figure 7-4).

Modularity of design is important for several reasons. First of all, if each function is physically modular, functions can be reused from product to product or from configuration to configuration. This reduces the number of unique designs required and can speed time to market. It also reduces the inventory of unique items.

The design on the left in Figure 7-4 has been haphazardly packaged. Each subassembly is unique and therefore more difficult to build (and to test). In the example on the right of the figure, function #1 has been partitioned to be modular.

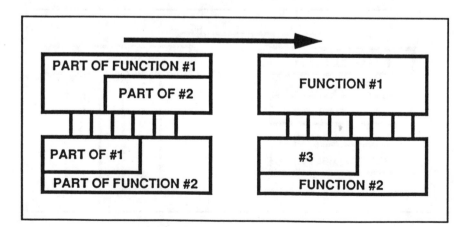

FIGURE 7-4   Physical modularity

It can connect to function #2, even if function #3 is added to function #2. *It is preferable to have multiple functions per physical module than to have one function split among several physical modules.*

Product weight is another very important design consideration. But it does not always pay to assume that less is best. The main thing to remember is that people will have to assemble, test, transport, use, and service the products that you design. The weight, therefore, must fit the application.

The general guidelines where weight is concerned include the following:

- Selected to fit the application
- Light enough in some cases
    - Man transportable
    - Aircraft applications
- Heavy enough in other cases
    - Telephones
    - Calculators
- Not usually significantly affected by DFM considerations

Think specifically about telephones that slide around on the desk during use—not heavy enough. In general, the application of any of the concurrent engineering guidelines, especially the design for manufacturability guidelines, have an insignificant (and sometimes even positive) impact on product weight.

Accessibility during system integration for calibration, testing, and rework are also important design considerations. Remember that human beings may have to duplicate and service the product you design multiple times. If it is difficult for you to do so once (during prototype assembly and debugging), think about the impact of that difficulty on the other organizations within the company over the life of the product. Where possible, *replace loose bolts and nuts with either captive hardware or quarter-turn fasteners for ease of access.*

CONNECTORS

Connectors are a major source of consternation in some operations. Each extra connection affects product reliability and cost. We wish to *use quick connect/quick disconnect* schemes in place of schemes that require soldering and unsoldering in almost every case.

Each connector should have as many pins as possible, within reason. This reduces the total number of connectors and provides extra pins for electrical access during testing. Connectors should be keyed to prevent improper mating (but keying should be defeatable for test purposes). Finally, connectors should be marked to help prevent incorrect connections, and the marking should have some semblance of order (i.e., P1 to J1, P2 to J2, and so on, and not the other way around!).

PRINTED CIRCUIT BOARD CONSIDERATIONS

Printed circuit board layout is becoming increasingly critical, especially as the use of surface mount technology (SMT) and fine pitch technology (FPT) continues to spread rapidly. There are several organizations that have put out guidelines on board layout. Two of them are the Institute for Interconnecting and Packaging Electronic Circuits (IPC) and the Surface Mount Technology Association (SMTA). Booklets from both organizations contain a great deal of detail regarding the board layout topics. We are going to take a general look at several printed circuit board considerations (with emphasis on the most important considerations).

First, *try to standardize* on one, or a few, basic printed circuit board (PCB) sizes. This reduces the costs of tooling and processing at the bare board vendor's factory, as well as the number of pallets and process variations required for soldering assembled PCBs in your factory.

Next, try to make sure that all PCBs are laid out using a *standard grid*. A 0.100" grid is preferred by most manufacturing engineers, as most existing equipment is designed to handle it. With SMT and FPT, obviously, spacings between component pins will be much less than 0.100". But many boards contain mixed technology that is amenable to 0.100" mounting of many components. Where component leads do not conform to standard grid locations, elongated traces can be used in the design to bring fixture-accessible points to a standard grid.

The point is: Do whatever you can to fit your new design into the existing process. *Consult with the manufacturing people* (and the test people regarding fixturing for test as well) early in the concurrent engineering design process, especially if your new design will require new assembly equipment or modified production processes.

Tooling holes are critically important for multilayer bare board fabrication and test and for automated assembly and test of loaded boards as well. As illustrated in Figure 7-5, there should be a minimum of three tooling holes provided on each board. These holes should be on a standard grid, should be a minimum of 1/8" in diameter, should be drilled (not punched), and should not be plated.

Punching, rather than drilling, the tooling holes results in significant inaccuracies. Even plating tolerances can cause test probes to miss test pads in FPT designs. If in doubt, consult the bare board fabricator and your manufacturing and test engineering groups. Component placement is also an issue, especially for dense designs. If components are placed too close together, soldering problems may

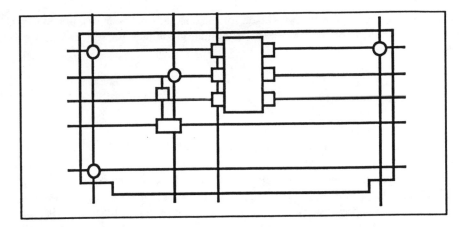

FIGURE 7-5   Tooling holes for printed circuit boards

occur. This can result in an increased number of open and short circuits that will required isolation and rework.

There is also an impact on the ability of automatic insertion and pick-and-place equipment to place components placed too closely together, especially if very small parts (e.g., chip resistors and capacitors) are placed next to large (or tall) components. Finally, functional board testers using IC clips (instead of single point probes) may not be able to use the multi-pin clips for fault isolation if components are spaced so tightly that the clips won't fit. *Leave as much space as possible between components.*

Space should be left around the outside borders of a board so that it can be palletized for soldering (and so that bed-of-nails test fixtures for manufacturing defects testers can be used reliably). In general, *0.200″* of empty space between board edges and components or traces is recommended. The absolute minimum is 0.100″ of clearance.

It is often more efficient and economical for a bare board producer to build multiple copies of individual boards on a large panel. Some loaded board manu-facturing operations also prefer to perform automatic component insertion on boards before they have been cut to their final size. Board size should take this into account wherever possible.

Each panel must, of course be inspected and tested after the entire sheet has been fabricated. *Proper panelization design* includes space between panels for a hole to be punched out beside any bad panels. Bad panels should also be marked (on both sides) with a large "X" in ink that will not wash off when the bare board is cleaned prior to delivery.

The need for *test pads* on printed circuit boards also impacts design for manufacturability. Space must be made for the test pads between components or on the bottom side of the board. The minimum guidelines for test pads are as follows:

- Aligned to 0.100" grid
- Minimum diameter of 0.040"
- At least 0.010" from nearest trace
- Ground test points every 4 square inches
- Through hole via or non-through-hole pad

Soldering, especially for SMT and FPT designs, is sometimes something of an art. It's possible, however, to reduce the "art" needed and make soldering more of a science. The successfulness of a soldering operation has a lot to do with the physical design of the printed circuit board. The direction that the traces run on a board in relation to the solder wave can have a big impact on the number of short circuits that occur as a result of the soldering process. This applies to both through-hole board designs and mixed through-hole and SMT board designs.

As illustrated in Figure 7-6, lines that run perpendicular to the solder wave will accrue fewer shorts than lines that run parallel to it. Long traces, in particular, and traces that are bus lines (with multiple components connected to them) should be laid out with the soldering process in mind.

Good soldering also requires plenty of clearance between component pads and via holes. We want to (a) avoid any shorts between pads and vias and (b) make sure that all vias get filled with solder. Proper clearance dimensions for both unconnec-

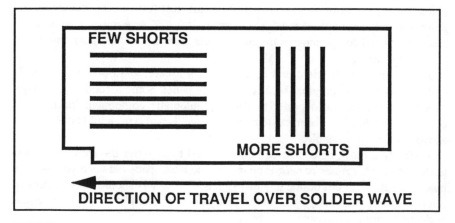

FIGURE 7-6   Trace directions for minimum solder shorts

ted and connected pads and vias are important, and IPC or SMTA guidelines should be followed.

There are also several guidelines for reducing manufacturing defects due to the soldering process and for making the soldering process itself easier. Since board edges get the hottest, its important to (a) keep small components away from them and (b) place large components, evenly spread out, nearer to the edges. Since large groups of large components will require more heating, they should be avoided. The heat required for large groups of large components for good soldering may cause excessive heat to be applied to smaller components (especially if they are near the board's edges).

Table 7-1 lists the clearances recommended between various types of components for good manufacturability of SMT and mixed SMT and through-hole designs. Note that, while chip resistors and capacitors may sometimes be placed as close as 0.050" to 0.060" (when adjacent to small outline ICs, for example), at other times a clearance of at least 0.200" (when adjacent to a large PLCC, for example) is required. The reason for the need for the extra clearance is to avoid having large components "shadow" the smaller components (which prevents them from being properly soldered).

The orientation of SMT components such as resistors, capacitors, and small outline transistors on the bottom side of the board is another important factor in achieving uniform soldering results. The *components must have adequate spacing* between them and be oriented so that the packages do not shadow the leads and so that no package shadows another.

Component identifiers should also be screened onto printed circuit boards

TABLE 7-1   Recommended component clearances

	RES OR CAP	SOT	SOIC - HORIZ	SOIC - VERT	PLCC
SOIC - HORIZ	0.060	0.060	0.100	0.100	0.200
SOIC - VERT	0.050	0.050	0.100	0.100	0.150
PLCC	0.150	0.150	0.200	0.150	0.300
AXIAL	0.050	0.050	0.150	0.100	0.150
DIP - VERT	0.125	0.125	0.100	0.125	0.250
DIP - HORIZ	0.125	0.125	0.100	0.125	0.250

whenever possible. Two possible methods for accomplishing this (see Figure 7-7) are:
- Screen the component identifier right next to the component.
- Use an "A, B, C...," "1, 2, 3..." type grid scheme for component location purposes.

ADJUSTMENTS

Adjustments complicate the manufacturing process, since they require in-process test steps and take extra time to accomplish. An adjustment, however, is preferred over a "select value at test" component. Select-at-test components automatically double the test time and result in secondary manufacturing operations. Sensitive adjustments should be designed so that they are not inadvertently knocked out of proper alignment. Adjustments that need to be performed frequently (why?) should be made accessible, and *interactive adjustments should always be avoided.*

BATTERY CONSIDERATIONS

Many computer products, among others, contain batteries for such things as real time clocks and random access memory (RAM) backup. It should be possible to replace batteries with power applied to the product in order to prevent the data the battery is designed to save from being lost when the old battery is removed. A dead battery should not cause the product to cease to function.

Battery life can be extended by providing a low power mode of operation (i.e.,

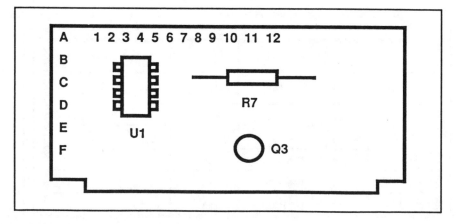

FIGURE 7-7   Component identifiers on boards

using software to shut down functional groups of components until they are needed). Rechargeable batteries or batteries that continuously "trickle charge" during product operation may be appropriate in some designs as well. Finally, remember to provide test points for testing the battery voltage independently of the product power supply voltage.

AUTOMATED ASSEMBLY CONSIDERATIONS

The selection of components during the design process can have a big impact on how well the product design fits into the manufacturing process. Sealed devices are preferred over open devices. The sealed devices are less susceptible to damage from either the soldering or cleaning operations. The actual component material also needs to be checked to make sure that it won't "melt" when cleaned.

Where automatic insertion equipment or pick-and-place machines will be used, components that can be supplied in tubes, taped and reeled, or in egg crate packages are preferred. Close coordination with vendors and your own organization's purchasing department is required in the area of component selection standardization. Components should be selected so that the number of variations in package sizes and styles are minimized. This reduces the cost of automated assembly equipment and cuts down on the time required for, as well as the complexity of, setup operations to change over for different package sizes and styles.

Be aware of the capabilities and limitations of the equipment available in the factory for automated assembly. Designs that require placement accuracies in excess of the capabilities of the manufacturing equipment will be prone to assembly defects. Components should be selected so that they can be placed and positioned with good repeatability.

There are several things to keep in mind during product design as the manufacturing operations learn to adapt to new competitive pressures. Remember that design decisions made early in the design phase of a product have a very large impact on overall life cycle costs. Smaller lot sizes will be the norm (with lot sizes of one being the goal in some organizations). Designs should facilitate flexible automation, lower inventory levels, and fewer part types from fewer vendors. The bottom line is more design standardization and more interaction with the manufacturing engineering members of the concurrent engineering product development team.

SUMMARY

The design for manufacturability guidelines that we've covered in this chapter can really be summed up as follows: *Don't design a square peg product to fit into a round hole production environment!*

Know the production environment. Take a factory tour or spend some time watching how your latest design is produced. You'll see what aspects of the design allow the product to flow smoothly through the assembly process and what aspects cause problems. With that knowledge, your next design can take advantage of the good things and minimize the not-so-good things.

It is not necessary to become an expert at manufacturing engineering to design a manufacturable product. It is necessary to be aware of the constraints—on layout, component selection, component placement, and assembly equipment—and to know who to consult in the organization for specific questions. If all of the constraints are identified up front, the proper engineering trade-offs can be made during the initial design. Then the design will fit into the process. If the constraints are not identified early, the result may be a costly and time-consuming redesign or a product that costs too much to produce to be competitive.

REVIEW

1. It is important to keep the total number of (a) *parts,* (b) *part types,* or (c) *both* to a minimum.
2. Assembly access from (a) *one direction* or (b) *multiple directions* is preferred.
3. It is preferable to (a) *split a function between multiple physical modules* or to (b) *place multiple modular functions on one physical module.*
4. If in doubt about the capabilities of the manufacturing processes, it is best to (a) *consult the manufacturing engineering team member before making a design decision* or (b) *make the design decision in a vacuum and hope for the best.*

Answers: 1: c, 2: a, 3: b, 4: a

NOTES:

8

Design for Testability Considerations

The increasing complexity of new products, and the proliferation of new electronic device fabrication and packaging technologies used to implement each succeeding new design, have made testability a necessary product performance attribute. For without testability, the most technically elegant product, from a "functions per square of space" standpoint, is absolutely useless.

It does absolutely no good to shave a day from design—ignoring testability—if that lack of testability adds weeks or even months to time to market. It must be possible to "design verify" and debug a new product design in the shortest possible time. It must also be possible to bring that product reliably to market in a competitive manner. This means that test programs must be generated to detect all of the possible faults that can occur in the product—both during product manufacturing and during the product's service life. It must also be possible to generate those test programs in a timely and efficient manner.

"Time to market" is a concept not always understood in the same way by different people. Some consider time to market to be raw schematic capture and design verification (e.g., good circuit simulation or prototype debug) time. Others realize that *true time to market is the time it takes from the beginning of design until the product can be successfully delivered into the customer's hands at a competitive price.* Figure 8-1 illustrates the effect that proper testability design can have on product time to market.

Design for testability is no longer just a "nice-to-have" feature in a product. Nor is it strictly an engineering discipline. It is, in fact, an element in a strategy of maintaining competitiveness in world markets, now and, especially, in the future. Companies can no longer continue to add cost to products through higher than necessary test programming times, test times, troubleshooting times, and high

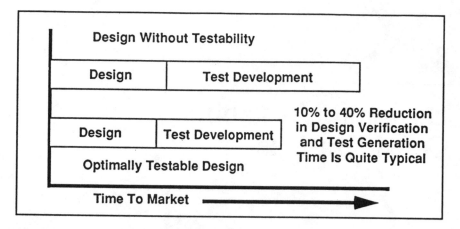

FIGURE 8-1   Testability impact on time to market

capital equipment costs. The technology now exists to build testing resources right into new product designs, thus drastically reducing the percentage of product cost made up by test and testing-related purchases and activities.

When one looks at the evolution in testing cost as a percentage of overall product (or overall business) costs, it becomes clear that something must be done about lowering those costs. Test does not add value to a product, it adds cost. Unless, of course, it is built-in test that provides a clear benefit to the customer as well as the producer. Examine the very simple and very real example shown in Table 8-1. The table illustrates the shift in parts, labor, and test costs that have

TABLE 8-1   Test cost as a percentage of product cost

	1986		1991	
	Actual	Percent	Actual	Percent
Parts Cost	$300	60%	$150	50%
Labor Cost	$100	20%	$50	17%
Test Cost	$100	20%	$100	33%
Total Cost	$500	100%	$300	100%

occurred over the last five years for a typical (e.g., 100 medium scale integration (MSI) integrated circuit (IC) equivalent), mostly digital board.

While parts costs have decreased due to improvements in product yields and increasing levels of integration, and while assembly labor costs have decreased due to the use of fewer numbers of components per board design and increased levels of assembly automation, test costs have not decreased at all. Thus, while total product cost has decreased by 40 percent (from $500 to $300), *test cost has risen as a percentage of product cost from 20 percent to 33 percent,* even though its absolute cost may not have changed.

In many cases, however, the situation is much worse than the example just shown. As complexity increases and new packaging technologies (e.g., multichip modules) continue to shrink parts count and labor content in new products, the actual dollar cost for testing has increased. This results in a test cost contribution to product cost in the range of 35 to 55 percent (or more!), depending upon product size, technology, and complexity. That is a lot of added cost.

How does one control the rapidly rising cost contribution of test- and testing-related activities—simply by making the unit under test (UUT) testable! Take as an example the revolution in computer technology. Computers continue to get smaller, more powerful, and less expensive. Testers, on the other hand, tend to get larger, incrementally less powerful, and far more expensive. Why is there such a dichotomy? Simply because the design of new products does not allow the testers to get smaller, more powerful, and less expensive! Most people are still using brute force (i.e., complex, expensive test equipment) to overcome unit-under-test-testability deficiencies. The key, however, lies in the prevention of testability problems. This is where the leverage is.

Testability is part of the concurrent engineering effort to reduce product costs and improve productivity and quality throughout the total business cycle—from product concept through design, manufacture, and usage in the field. Testability is not an attempt to restrict engineering innovations or to criticize the ability of the design (or the designer!) to perform its function. We just want the ability to test the best (from a functional point of view) products in the least amount of time and at the least cost. This means that future products must be testable!

HOW AND WHY CIRCUITS ARE TESTED

To understand the importance of testability, it is necessary to understand how circuits are tested and why circuits are tested. Circuits are tested by applying stimulus signals, either digital or analog in nature, to circuit input pins, and verifying the response of the unit under test to those input stimulus signals by evaluating the response signals supplied by the unit under test. In the digital realm, the input stimulus signals, usually called test vectors, and the resulting response signals are patterns of logic 1s and logic 0s. In the analog world, the input stimulus signals, and thus the resulting responses to be analyzed, can be quite complex variations of frequency, voltage, resistance, or other parameters.

In any case, the reason for applying and analyzing test stimulus and response signals is to detect all of the possible faults in a unit under test that could prevent proper circuit operation. Faults can occur in the components used in an assembly or as a result of the assembly process. Faults that occur within the components are usually referred to as functional faults; those that occur on the assembly are usually referred to as manufacturing defects. Faults that occur due to functional interaction problems between good components on a good assembly are design defects; those that occur after an assembly or system has been placed in service are usually functional faults.

Regardless of the source of the fault, its effect is to prevent proper circuit operation. The test process must, therefore, detect all of the possible faults that could prevent proper circuit operation. In the digital realm, this is accomplished by applying stimulus vectors (i.e., sets of logic 1s and 0s) to circuit input pins in an attempt to cause every node (i.e., circuit interconnection) in the circuit to be at the logic 0 state at least once and to be at the logic 1 state at least once. This is called fault activation. Activating faults, however, is not enough.

Faults must also be propagated to circuit output pins in order to make sure that each node actually assumed the state that it was ordered to assume by the stimulus vectors. Faults are propagated by providing a path from the circuit node being activated through the other circuits in the design to circuit physical output pins (or test points). Only when a fault can be both activated and propagated can it be detected. Thus, fault detection requires activating faults and propagating faults.

The analog world uses a similar methodology but different input signals and output response analysis methods. Analog stimulus signals are applied to circuit input pins in order to exercise each circuit node to its full range of parameters. This is analogous to the digital logic 0 (low) and logic 1 (high) states. Analog responses are measured parametrically (i.e., volts, amps, time, frequency, and waveform characteristics) in order to determine if the node under test actually did what it was supposed to do.

The concept of fault activation and fault propagation is illustrated in Figures 8-2 and 8-3, using a digital circuit example. In Figure 8-2, the objective is to answer the question "Is IC5's output stuck at the logic 1 state?" In order to answer that question, it is necessary to place test signals on the inputs to cause IC5's output to go to the logic 0 state (if it is working).

In the example of Figure 8-2, IC5's output will go to the logic 0 state when both of its inputs are sent to the logic 1 state. To provide a logic 1 state at the output of IC1, its inputs must both be at the logic 1 state. To provide a logic 1 at the output of IC2, both of its inputs must be in the logic 0 state. Thus, the two logic 1s and the

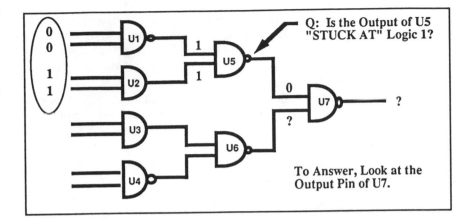

FIGURE 8-2   Fault activation

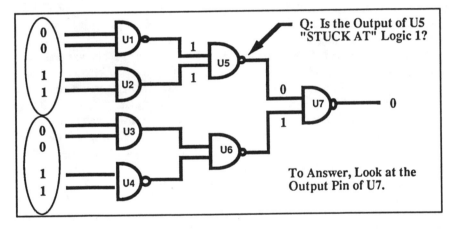

FIGURE 8-3   Fault propagation

two logic 0s at the upper left-hand side of the figure make up the portion of a test stimulus vector that will activate the fault "IC5 STUCK AT 1." In other words, if IC5's output is not "stuck" in the logic 1 state (i.e., it is not faulty), it will assume the logic 0 state with the test stimulus vector so far applied.

The testing job is not, however, complete. Even though the initial test vector applied so far has created a set of conditions that will cause IC5's output to change to the logic 0 state in response to a test stimulus vector, there is currently no way of observing whether or not IC5's output actually went to the logic 0 state. There is no known path through IC7 to allow observation of the results of the input stimulus to U5.

Logic 1s and 0s must be applied to the inputs of IC3 and IC4 in order to provide the correct states to IC6 to send its output to the logic 1 state as well. With IC6's output at logic 1, IC7 will transfer the results of activity on IC5's output to the circuit output pin. The test vector is now complete, and we can detect a "stuck at 1″ fault at the output of IC5.

This test vector, in fact, detects quite a few faults in addition to IC5 stuck at 1. It detects input faults stuck at 0 for IC1 and IC3, input faults stuck at 1 for IC2 and IC4, output faults stuck at 0 for IC1–IC4, input stuck at 0 faults for IC5 and IC6, output stuck at 0 for IC6, the top input of IC7 stuck at 1 and the bottom input of IC7 stuck at 0, and the output of IC7 stuck at 1. If IC7's output does not respond correctly to the input test vector, where is the fault?

That question brings up the concept of "ambiguity groups"—if there is a fault, which component is actually causing the fault? Diagnostic accuracy may not have been important when the wrong part being replaced cost only a few dollars, but it is critically important with more expensive parts such as application specific ICs (ASICs), microprocessors, and other complex very large scale integration (VLSI) devices, and as rework costs continue to increase, especially for surface-mounted components and die-mounted-to-multichip module substrates.

The example presented above is admittedly somewhat trivial, but was presented to get certain concepts across to the reader. Consider what happens with a much more complex circuit such as the one shown in Figure 8-4. This circuit is composed of 20 ICs, some VLSI ICs and some ASICs, along with regular "glue logic." It may contain feedback loops, it may or may not be initializable, there may be long data paths or counter chains between the physical inputs and outputs, and there are many levels of logic to be stimulated in order to both activate and propagate all of the possible faults that could prevent proper circuit operation. This circuit is extremely difficult to test unless testability features are added.

Consider only the task of trying to find one of the thousands of possible faults in this circuit—that of IC U3's output stuck at one (or zero). It may or may not be too difficult to activate the fault by stimulating U1 according to its truth table or functional behavior. Depending, of course, on whether what is done to U3 affects U4 in such a way as to feed a signal back to U1 that conflicts with the desired applied stimulus vectors. But, even if the feedback loop does not present a problem, propagating that fault to the circuit output (at the output of IC U20) requires that all of the other circuitry also be stimulated. It is no longer a trivial task.

BASIC TESTABILITY AXIOMS

In order to make today's (and tomorrow's) complex designs testable, three key testability principles must be included in the circuit design. These three key testability principles, implementable in a great many ways, make the difference

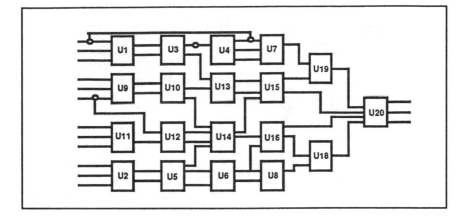

FIGURE 8-4   A more complex example

between an inherently testable design and an untestable one. These three key testability principles are:

- Partitioning
- Controllability
- Visibility

Circuits are partitioned by breaking them into *reasonably small functional blocks,* or clusters. This makes them easier to understand, easier to write tests for, and easier to test and troubleshoot. Circuit control is provided by including *reasonably direct paths* from the test resource (either automatic test equipment or built-in test circuitry) to critical internal nodes required for initialization of the circuitry under test, partitioning of that circuitry, and control for fault activation. Circuit visibility is implemented by bringing *internal nodes to the testing interface*, again in a reasonably direct manner. This principle reduces logic and fault simulation times and costs along with test generation, testing, and troubleshooting times.

These three key principles are easy to implement with minimal effect on circuit configuration, performance, and reliability. The circuit of Figure 8-5 shows four gates added to the complex example to partition it into smaller functional blocks. This allows certain blocks of the circuitry to be stimulated while leaving other circuit elements in the inactive state (i.e., not needing to be stimulated to activate faults in the subsequent circuitry).

The added gates shown in Figure 8-5 may, in some cases, add an extra delay between circuit elements. In other cases, no delay is added. The circuit is simply modified to include an extra input (called an extra fan-in point) to directly control

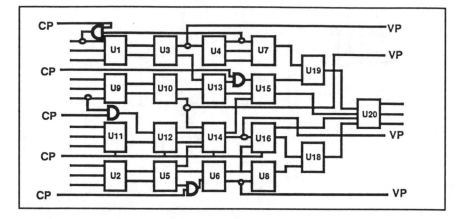

FIGURE 8-5   Adding partitioning, control, and visibility

a circuit function for testing purposes. This extra input drastically cuts the time needed to generate the stimulus vectors necessary to activate faults.

Large sequential circuits are also not always initializable (i.e., able to be set immediately to known states for testing) and often contain feedback loops and long counter chains. A few strategically selected control points allow for direct (immediate) initialization of memory elements and other sequential circuits. They can also be used, usually with an extra gate or two, to break feedback loops. Examples of these types of test control points are also shown in Figure 8-5.

A "reset" line has been added to provide immediate initialization, and an extra input has been provided to allow the feedback loop to be disabled, eliminating any conflicts between the desired applied stimulus patterns at the inputs to IC U1 and the results of those patterns on the outputs of U4. Lack of initialization and the presence of feedback loops contribute significantly to long test generation, logic simulation, and fault detection times (and resulting higher costs).

To further reduce times and costs, visibility (or observability) points are added so that extra stimulus patterns are not needed in order to propagate faults to circuit output pins via other complex circuits. With visibility points added, faults are propagated immediately to the tester or built-in test interface. Figure 8-5 further shows the addition of four visibility points to the already partitioned and controllable complex example.

Consider now, with the addition of four simple gates, five input control points and four output visibility points—the problem of activating and propagating the fault "U3 stuck at one (or zero)." It is an infinitely simpler problem to solve now that the circuit has been partitioned and made controllable and visible.

IC U3 can be directly stimulated through U1 without worrying about the effect of U3's outputs on U4. The actual output of U3 can be directly and immediately observed by the tester or built-in test circuitry without stimulating all of the surrounding or intervening circuitry. In short, these *three key testability principles—partitioning, control, and visibility*—have a dramatic positive effect in lowering all times and costs associated with both design verification and test. Also, as shown, their implementation requires very little in terms of extra circuitry or extra circuit input/output connections.

Testability techniques are designed to reduce, or at least control, ever-escalating test costs. Product design engineering team members today must design circuits that can be tested in an efficient, economical, and orderly manner. Incorporating features that facilitate testing and fault isolation and that help to reduce maintenance costs over the life cycle of the product must become part of the job description of the product designer and part of the specification for each new product.

Customers should require testable designs from their suppliers. Producers should look at testability not as a burden to their product designs but as a feature that will make the product more attractive to the potential user. Design teams must make sure that every product design, however technically elegant, can also be produced, tested, and serviced at a competitive cost. If it is not, it won't be as successful as it could be.

TESTABILITY IMPLEMENTATION ALTERNATIVES

The "ad hoc" testability techniques used in the example just covered are examples of some of the simplest solutions to many major testability problems at the printed circuit board level. There are other constraints and techniques, however, that can make device level and system level testability implementation a little more complicated, or a little more sophisticated (depending on your point of view), than the method just described.

Some of these other alternatives include:

- Internal serial scan approaches
- External scan approaches
- Probe-ability approaches
- Testability bus approaches
- Combinations of the above

Internal serial scan approaches are most often used at the semiconductor device (or chip) level to make internal device nodes easier to control and observe. Scan designs replace all sequential circuit elements with structured sequential elements that can be connected in a serial chain and accessed via just a few—typically four—device package pins. They require anywhere from 12 percent to 20 percent of available silicon area, but can facilitate automatic test pattern generation and increase fault coverage (i.e., test quality).

External scan approaches are aimed at verifying board and, sometimes, system level structural (i.e., interconnect and functional) integrity. The most widely recognized external scan approach is boundary scan, which is discussed in a later section of this chapter. So-called probe-ability approaches use mechanical guidelines, such as those discussed in Chapter 7, and so facilitate probing of all (or at least most) circuit nodes to avoid as much as possible the need for including electrical testability circuitry in a design.

The IEEE's Computer Society has a group working on several versions of P1149.x testability bus standards. IEEE-Std-1149.1, the boundary scan standard, is the first to be formally approved. P1149.2, an I/O scan approach using a combinational, rather than serial-only, test access port is well on its way to approval, as is P1149.3, the direct access real time (addressable) testability interface. Work is beginning again on P1149.4, the analog testability bus standard, and P1149.5, a backplane bus modeled after the TM-bus developed for the very high speed integrated circuit (VHSIC) program for the U.S. Department of Defense, is moving toward IEEE approval.

Each implementation technique alternative has its own advantages, disadvantages, and best applications. In a concurrent engineering environment, the product birthing team must make quantitative trade-offs to determine which testability approach is best. (Note: For a complete discussion of testability implementation techniques, refer to the book Design To Test—2nd Edition, ISBN 0-442-00170-3, by Jon Turino, Van Nostrand Reinhold, 1990.)

BUILT-IN TEST CONSIDERATIONS

Integrated built-in test is clearly the way of the future in modern electronic design for all but the simplest consumer products, and even some of them (i.e., dishwashers, refrigerators, and automobiles) warrant its inclusion for increased customer satisfaction and lower manufacturing and service costs.

Built-in test implies both internal stimulus generation and response evaluation capabilities. Built-in test (BIT) and built-in self-test (BIST) are two phrases sometimes used interchangeably to describe the same thing. BIST is usually used when speaking of chip level design. BIT is more generically used at all levels of design.

In any case, as can be seen from the diagram of Figure 8-6, the implementation of built-in test can have a significant impact on the unit under test. Concurrent

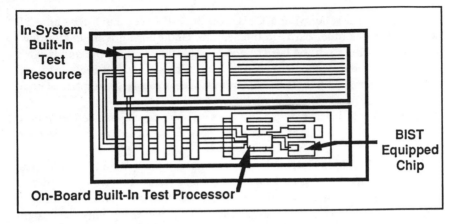

FIGURE 8-6   Integrated built-in test

engineering is an absolute must if BIT is to be implemented with minimum impact on system hardware and software overheads.

The key thing to remember about implementing built-in test is that it must be considered *during the design phase of the product development cycle.* Trying to "retrofit" built-in test or any of the other designs for testability guidelines into a design is often an expensive, time-consuming, and frustrating task.

There are several variations on the theme of built-in test. Three of these variations include:

- Centralized built-in test
- Distributed built-in test
- Remote diagnostic built-in test

If we look at the **centralized local built-in test** option, we see the following characteristics:

- It uses the system CPU as an ATE for testing.
- It requires access to most PCB edge connector signals.
- It can provide diagnosis to the board level.
- It is usually a large software development job.
- It provides no help for PCB component level diagnostics.

All of the built-in test code using centralized built-in test is typically resident in (or downloaded to) the system CPU, which really acts much like an ATE system to test the other cards in the system. Good fault coverage from just the system bus lines will be hard to achieve, particularly for the dumb cards (i.e., those without on-board processors and the associated read-only and random access memory resources), without access to a large number of edge connector signals.

Centralized built-in test can also be a large software development effort. It may, however, be the only option if all (or even most) of the boards in the system are dumb cards.

Distributed built-in test, on the other hand, uses the system CPU mainly as an initiator and monitor for built-in test programs resident on (or down-loaded to) each of the boards in the overall system. Depending upon the design of each board, a high level of fault coverage may be achieved using only the system bus pins—presuming the system bus is operational.

Distributed built-in test partitions the software development job and can be used as an aid for component level diagnosis on the card. It does, however, require that all boards have processors on them. Since that is not always the case, what is usually required is a combination of centralized and distributed BIT implementations.

Regardless of whether a centralized or distributed BIT approach is taken in the system itself, another option that exists is the option for remote diagnostic built-in test. With **remote diagnostic BIT,** the central processor is augmented with a communications device (i.e., a modem) that can communicate directly with an off-site diagnostic resource. This approach is quite common in the computer and telecommunication industries already. The main advantage to remote diagnostic BIT is that we can ship replacement assemblies to customer sites, rather than having to ship people with a whole array of replacement assemblies.

It is important to note that BIT design takes a different "mind set" than normal system design. Not only do we need to understand that go/no-go–type BIT is only useful when the system is working, we also need to design the BIT so that it can function and diagnose faults when the system itself is not working. That means we must consider a lot of "what if...?" scenarios. The usual method for achieving BIT success when the system ceases to function is to *provide an alternative signal path* into each major subassembly in order to reconfigure it and test it for diagnostic purposes.

The drawing in Figure 8-7 summarizes this discussion of BIT architectures, alternatives, and implementation. It's designed to show how the BIT design, allocated to the various levels of integration of the final system (i.e., product), is a *concurrent process, simultaneously involving the system, hardware, software test, and service design functions.*

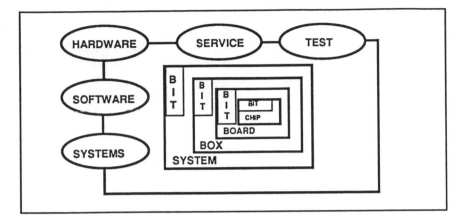

FIGURE 8-7   Built-in test summary

SURFACE MOUNT TECHNOLOGY
CONSIDERATIONS

With silicon integration improvements have also come changes in packaging technology. While surface mount and fine pitch technologies have been used in hybrid circuits and by the military services for many years, their mass commercialization is a relatively new phenomenon. SMT and FPT designs make it more difficult to use traditional testing approaches because of higher package densities and closer lead spacings.

It is especially apparent that fine pitch designs (including those implemented with multi-chip modules), which are clearly going to be the mainstream technology of the 1990s, will severely strain the ability of the in-circuit tester with its bed-of-nails fixture. While the mechanical guidelines—keeping the board reasonably small, keeping tall components away from test pads, making sure that there are enough (and large enough) test pads, and providing space around components to facilitate probing—focus on making FPT designs "probe-able," at least to some extent, they may be difficult to implement when the reasons for implementing FPT designs are examined.

Many people are beginning to find the electrical guidelines—testable devices, using multiplexers and shift registers to increase controllability and observability, and including more built-in test—more palatable. In high-yield situations (which must be the norm in the future for competitive electronics companies), only those boards that fail to work will be placed on ATE systems for fault diagnosis and isolation. If the board designs include the right kinds of diagnostic aids, the diagnosis of the few faults that do occur will be quick and correct the first time.

BOUNDARY SCAN

The idea for boundary scan originated in Europe with Philips, Siemens, Thompson and British Telecom (among others). Philips, as the organizer of a group called the European Joint Test Action Group (EJTAG), was particularly concerned with the first four items on the following list and wanted to develop an alternative to bed-of-nails test fixturing requirements.

- SMT and FPT will make the bed-of-nails fixture impractical in newer, more complex designs.
- Opens, shorts, and other "manufacturing defects" will continue to be encountered.
- Functional test programming is too expensive.
- Guided probing is too slow and cumbersome.
- On-chip BIST will become more prevalent.

As more companies (and countries) got involved in the boundary scan work, the "E" was dropped from the acronym and the work continued under the auspices of JTAG, the Joint Test Action Group. This group has grown widely since its inception in 1985 and continues its work to pressure all merchant (and ASIC) semiconductor suppliers to include boundary scan in all integrated circuits. Its technical committee has been added to the IEEE P1149.X family of Working Groups.

Several companies have introduced products that include, interface to, or otherwise support the boundary scan standard. Texas Instruments, AT&T, Alpine Image Systems, National Semiconductor, and Logical Solutions Technology Incorporated were among the first to support the standard, with others following rapidly. Intel and Motorola have already introduced leading edge processors that support the standard as well, and most ASIC vendors include the cells in their libraries. CAE and ATE tool vendors are also introducing products to support boundary scan.

While boundary scan is a good tool in the testability toolkit, one hopes that some of the reasons cited above for its invention will go away through the continuous improvement of both the design and manufacturing processes. Opens, shorts, and other manufacturing defects *must be prevented at their source*. The "factory within the factory," whose job is to repair defects, must be eliminated if manufacturing operations are to be truly competitive. Functional test programming is not too expensive when real electrical testability is implemented as part of the concurrent engineering process. And guided probing is too slow and cumbersome only in the face of large quantities of faults.

The illustration in Figure 8-8 shows an IC containing boundary scan cells and how the cells are connected together. Signals at the top of the IC are normal input signals. Signals at the bottom of the IC are normal output signals. The signals on the left are the Test Data Input (TDI), Test Mode Select (TMS), Test Clock (TCK),

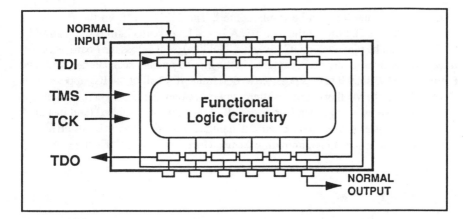

FIGURE 8-8   A boundary scan device

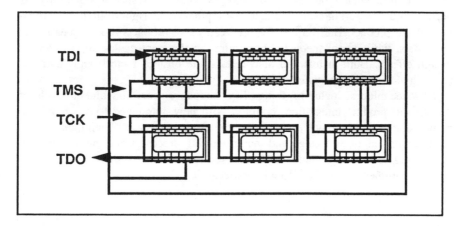

FIGURE 8-9   A boundary scan board

and Test Data Output (TDO). These signals come from a test resource of some sort (i.e., an ATE system, a built-in test resource or an on-board test access port interface device) and control the flow of data through the boundary scan chain.

Data coming into the IC can be captured in the boundary scan cells and clocked out on the TDO line. Data can also be clocked into the boundary scan cells on the TDI line and output on the IC pins for transfer to other devices. Figure 8-9 shows how multiple boundary scan chips can be interconnected at the board level to facilitate testing for manufacturing (and some other) defects.

When multiple boundary scan chips are assembled on a board, a four-wire testability bus interface can be implemented. The TDO line from the first IC is connected to the TDI line of the next IC. A board level scan chain is the result.

Signals can be sent into one IC (over the TDI line), sent across the board's copper traces connections to other ICs, captured in the input boundary scan latches, and sent out via the TDO line. It is thus possible to determine if (a) all of the ICs appear to be alive and well, (b) whether or not the interconnect between ICs is correct (i.e., there are no shorts or opens) and (c) whether or not all of the parts are installed (and installed correctly). In essence, most of the common manufacturing defects can be tested for via the boundary scan path, thus eliminating the need for the bed-of-nails fixture.

SUMMARY

Design for testability of electronic products is a large and well-documented part of the concurrent engineering discipline. Because of the application specific nature of testability implementation, test engineering personnel with specific testability expertise should be part of every product development team right from the specification of the product. Built-in test features can make products more appealing to customers, while simultaneously shortening development cycles due to parallel design efforts. Built-in test can also reduce the cost of factory and field test equipment and their associated operations.

New techniques, including things like boundary scan, and higher levels of device and product integration, coupled with new automated tools, make implementing testability much easier today than it was years ago. Testing operations are the monitors of the design and manufacturing processes and the information that testing operations provide should be used for continuous improvement of the design, manufacturing, and testing operations.

REVIEW

1. When design for testability features are implemented in new product designs that require software design verification, including logic and fault simulation, attention to testability during the product design phase will (a) *increase* or (b) *decrease* the time required to prepare a design for manufacturing and test.
2. Testing during manufacturing is designed to (a) *continuously reverify design parameters* or (b) *make sure that a presumably good design was duplicated properly*.
3. Testing requires the (a) *activation or faults*, (b) *the propagation of faults*, or (c) *the simultaneous activation and propagation of faults*.
4. Design for testability axioms include (a) *partitioning*, (b) *controllability*, (c) *visibility*, or (d) all three.

5. Built-in test is most effective when it is (a) *added to an existing design* or (b) *designed into a new product from the start.*
6. Boundary scan is intended as a replacement for (a) *in-circuit testing* or (b) *functional testing.*

Answers: 1: b, 2: b, 3: c, 4: d, 5: b, 6: a

NOTES:

9

Design for Serviceability Considerations

Many of the techniques used to improve the manufacturability and the testability of a product also serve to improve its serviceability. There are, however, some additional considerations that must be taken into account as part of the concurrent engineering process that are fairly specific to service issues. Some of the technical guidelines for serviceability may conflict with some of the guidelines for manufacturability, and this is perfectly okay.

Product design using the concurrent engineering process is often a series of constant conflicts and trade-offs that must be resolved through quantitative analysis, keeping the needs of the customer always foremost in mind.

ELEMENTS OF SERVICEABILITY

There are six major elements, or categories, of serviceability. These six elements are:

- Diagnoseability
- Accessibility
- Replaceability
- Repairability
- Rebuildability
- Upgradeability

Diagnoseability is defined as the ease with which the cause of a problem can be determined and the speed with which the failed part can be identified without expensive test equipment. One of the things that we do to improve diagnoseability

114

is to include built-in test features that eliminate (or at least minimize) the need for expensive test equipment. The electrical and physical partitioning that we've included in the design also has a large effect on how fast and easy (or slow and difficult) it is to diagnose a fault to a particular replaceable item.

Accessibility is the ability to work on systems, subsystems, and boards without interference from other unrelated components. Thinking about accessibility requires thinking about which items in a product are most likely to fail and making them the most accessible. We want to be able to remove individual replaceable items without having to remove other items first. If we remember the "nested" or top-down design for manufacturability guideline from Chapter 7, that may mean trading off some manufacturability for some serviceability (or vice versa). In other cases, the two guidelines may be compatible. The key is to think about the trade-offs during the design phase of the product.

Replaceability is the ease with which a part or subassembly may be replaced in the least amount of time with the least amount of skill with the simplest tools. We've hopefully helped to achieve that goal by using snap-together construction wherever possible (since it minimizes the number and types of tools required). We'll also want to make sure that assemblies connect together without soldered interconnections, since unsoldering and soldering an item in order to replace it is not a quick operation, requires certain soldering skills, and takes a non-simple set of tools.

Repairability is the ease with which a board or other subassembly can be fixed, rather than having to be replaced as an assembly and returned to a depot or other service operation. Repairability is yet another variation on the serviceability theme. Here we're talking not about replacing a failed item but actually repairing it in the field. Techniques such as socketing major components can be used to enhance the repairability of an item. The service philosophy and organization of the business will have a big impact on the level of repairability required in a product. If, for example, board swap is the preferred maintenance philosophy, field repairability of boards is a much lesser concern than it would be if the strategy was on-site repair to the component level.

Rebuildability is the ability to salvage costly components by rebuilding, rather than scrapping, expensive subassemblies. The objective in making something rebuildable is simply to lower scrap costs. If a design requires riveting or gluing components to a board, for example, its rebuildability quotient may be quite low (since removing a failed glued down or riveted component may cause damage to the board that cannot be repaired). If, on the other hand, it will be less expensive to simply discard a failed item, rather than to try to rebuild it, rebuildability has a very low priority in the product development process.

Upgradeability is the ease with which products can be modified or expanded to include new or improved features in the field environment. In this day and age of programmable products, where both hardware and software may have to be changed once the product is in the customer's facility, upgradeability can be an important factor. Easy upgradeability can be a powerful marketing tool as well as an internal tool to prevent product obsolescence and allow for the reuse of product modules.

ACCESSIBILITY REQUIREMENTS

Depending upon product size, use, and configuration, the following guidelines apply to providing for accessibility:

- Place subassemblies on slides.
- Provide service loops for cables.
- Make test points accessible.
- Provide provision for PCB extender cards.
- Provide strain reliefs on cables on subassemblies.

If we build large systems with multiple subassemblies, we can put those **subassemblies on slides** so that they can be extended for service purposes. We want to use the kind of slide assemblies that have locking mechanisms, as well, so that we don't pull an assembly all the way out and have it fall to the floor (which could damage both the system's cables and the service person's bodily parts!).

When we do put subassemblies on slides, we need to remember to **provide service loops on cables** so that power can be applied to the subassembly for troubleshooting purposes. This has electrical ramifications as well. The circuitry must be designed to handle the extra wire lengths (including enough noise immunity to overcome any crosstalk problems created by the longer cables). Jumper cables can sometimes be substituted for service loops, but they tend to get lost soon after the system is delivered!

Test points, or at least critical or often used test points, should be accessible without removing circuit cards from backplanes or card cages whenever possible. These test points should be chosen to allow us to determine whether a board is faulty and not necessarily to determine which component on the board is defective. For that we'll have to remove the board and place it on an extender card, as illustrated in Figure 9-1.

When **providing provisions for extender cards,** we need to keep the size and weight of the board under test under control, as well, so that it will not pull itself free from the extender card during the troubleshooting process. It's also nice to put mechanical test point connectors of some sort on the extender card to facilitate the connection of test equipment.

Cables should be provided with **strain relief mechanisms** so that wires do not

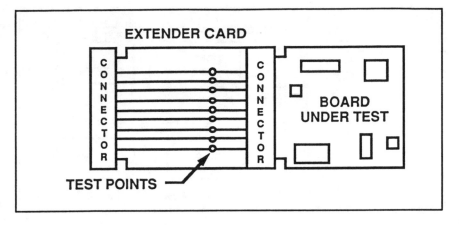

FIGURE 9-1   Providing provisions for extender cards

get broken when cables need to be disconnected. It is bad practice to remove cables by pulling on the actual wires, so the connector itself should be designed to be easily removable by grasping it (and not the wires that are installed in it).

The actual connectors used, keeping in mind the design for manufacturability guideline that says to minimize the number of cables (and therefore connectors), should be positioned properly so that they can be removed without removing other subassemblies whenever possible and without tools whenever possible. They should also be robust enough to handle multiple connection/disconnection operations without damage so that they can support multiple service operations.

External connectors should have positive seating, or locking, mechanisms so that users know definitively when they are correctly installed. Recessed contacts should be used and are important in preventing damage from inadvertent shorting (or danger to the service person from high voltages). Polarization or keying helps prevent connectors from being inserted wrongly (which can cause significant damage to the product).

FUNCTIONAL MODULARITY

Functional modularity was discussed in Chapter 7 under the topic of physical partitioning. It is important enough to merit further discussion. In a large system, we want each replaceable element to represent a complete function (rather than having a function spread over multiple replaceable elements). On a processor board such as the one illustrated in Figure 9-2, we want to be able, for example, to replace the ROM with a newer version or to add expansion RAMs in the field. The design must take these requirements into consideration.

Additional functional modularity guidelines to make servicing products easier

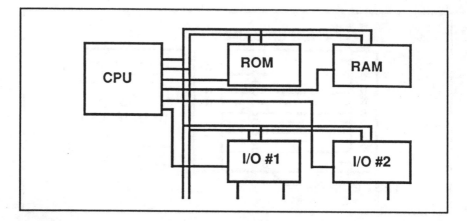

FIGURE 9-2   Functional modularity

apply to such things as power supplies. High stress components (such as flyback transformers in video terminals, for example) should be made accessible for quick and easy replacement. Space should be provided around fuses to make them accessible, if appropriate, without disassembling the unit under test. Fuses can often be replaced with circuit breakers to minimize the need for replacement parts.

In high voltage power supplies (or other high voltage circuits), bleeder resistors and shorting straps or bars should be provided for discharging high voltage nodes. Adjustments should also be placed as far from hazardous voltages as possible.

In general, try to minimize the number and types of tools required to service the product. Use as few screws as possible and make them captive where possible. Snap-and-clip and latch-and-snap fastenings are preferred to nut and bolt fastenings. Just remember that they must be durable enough to handle repeated operations and robust enough to resist damage from repeated assembly and disassembly. Let's also remember the user of the product! When it comes to controls, the ones that are used frequently should be easily accessible to the operator from the outside of the enclosure and their functions clearly identified.

External controls are best when they cover small ranges (i.e., fine adjustments). Internal controls can be set to adjust for circuit tolerances (i.e., coarse adjustments) and to center the range for the external controls. Lastly, try to use adjustable components that can all be adjusted with the same tool (or small set of tools). Several physical packaging and assembly methods can be used to enhance the serviceability of a product. Three are illustrated in Figure 9-3.

Assemblies should be made pluggable whenever possible. Extra circuitry for optional features can sometimes be housed on "daughter boards," and ICs on boards can sometimes be socketed. This might be particularly advantageous, for example, when upgradeability is a critical factor and new read-only memory

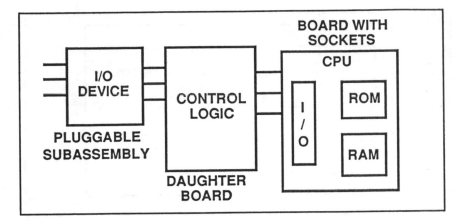

FIGURE 9-3   Physical partitioning

devices might need to be field installed, to complete a software (or firmware) upgrade.

Not all of these techniques can always be used all the time. High reliability requirements, for example, including the ability to operate under shock and vibration conditions, may preclude the use of IC sockets. But for many products, these techniques are just fine.

LIVE SYSTEM BOARD REPLACEMENT

Some systems, most notably fault-tolerant computer systems, must be designed so that they will continue to operate even while being serviced. This requires the ability to replace boards without (a) glitching the system or (b) damaging the board (or backplane connector) while the power is still on.

In addition to the normal requirements for redundancy, built-in diagnostics, and dual bus operation, there are some techniques for designing boards so that live system replacement is safer and more reliable. When it comes to live system board replacement, if all of the traces on a board edge connector are the same length, then theoretically all of the pins will make contact with the connector pins at the same time. In reality, however, there is always some difference (usually on the order of milliseconds) between when the power and ground lines make the connection.

In order to make sure that the +5 volt power rail stays above ground and that the -5 volt power rail stay below ground, a pair of Schottky diodes can be added, as shown in Figure 9-4.

Plugging in a card with large value capacitors on it can cause the system power supply voltages to dip momentarily, thus upsetting the operation of the system. One way to solve this problem is to stagger the traces on a card so that the ground trace

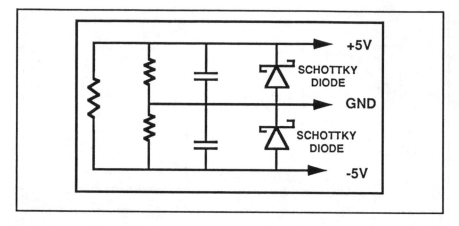

FIGURE 9-4   Diode protection

and a trace connected to a card pre-charge power supply make contact first. This way the capacitors on the board can be pre-charged before the main +5 volt trace makes its connection.

There are right and wrong ways to design the card edge and the power traces for live board insertion. Some examples of both are shown in Figure 9-5. The average time delta for making contact is about 1 millisecond for each 0.010 inches of trace length, but this can vary (up or down) by a factor of 2, depending on insertion force. The key to proper edge connector design is to make the timing of the fingers as independent as possible of physical forces (such as the angle of insertion or the insertion force).

Proper electrical and mechanical design for live board replacement means making sure that the pre-charge fingers (if used) make contact well before the main +5V (or other power supply voltages) and ground fingers make contact. Series diodes can also be placed in any card pre-charge lines to isolate the pre-charge supplies from the main power supplies once the card has been inserted.

ANALOG AND HYBRID CIRCUIT CONSIDERATIONS

Many products, even if they are mostly digital in nature, often contain some small percentage of analog circuitry. Analog circuitry can be tougher to test and service than digital circuitry due to the larger number of discrete components usually associated with it and because of the wider ranges of circuit parameters normally encountered.

Dealing with very small signal levels, either for measurement or stimulus purposes, is often a difficult task requiring highly accurate test equipment with

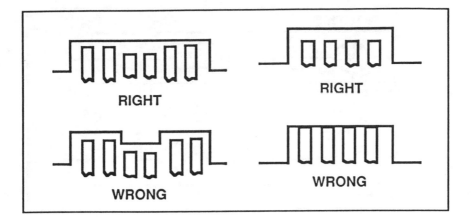

FIGURE 9-5   Circuit card edge connector design

high resolution. The need for sophisticated test equipment can be reduced and the accuracy of measurements increased if signal levels are kept fairly large. Sending larger signals across interfaces also improves operational reliability by improving noise immunity.

Remember also that not all service personnel are as consistent as you are when it comes to turning off the power. Try to partition designs so that removing one card will not cause other cards to malfunction or be damaged. And as for test points in discrete analog circuits, a good place to pick up a test point is at the top of an emitter resistor. Even if there is a bypass capacitor across the emitter resistor (which would prevent us from seeing any AC waveform at the test point), the DC voltage at the test point can give us a good indication of whether the transistor providing the gain in, for example, an amplifier stage is open, shorted, or operating in the normal range.

Partitioning is just as important for analog circuits as it is for digital circuits. Figure 9-6 shows an example of a mixer that includes two amplifiers and a filter. If the filter requires alignment, the presence of the removable jumper, as shown at the bottom of the figure, will make the process far simpler and require less variety of test equipment.

Test points at the outputs of analog circuits that drive electromechanical components (like meters or displays) allow the circuit parameters to be measured without operator intervention and should also be included in circuit designs. It may be necessary in high impedance or high frequency circuits to buffer the signals before sending them out to a physical test point. This may require an additional operational amplifier configured as a voltage follower or a discrete transistor amplifier configured as an emitter follower.

Where high voltages must be monitored, use a voltage divider to reduce the

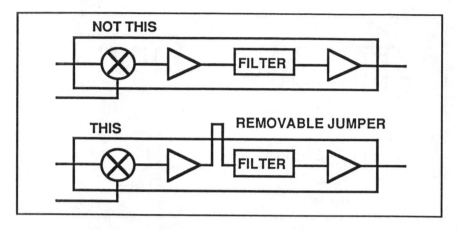

FIGURE 9-6   Analog partitioning

voltage that is actually sent out to the test point. It is good practice to put two resistors, connected in parallel, in the bottom leg of a voltage divider. This way if one of the resistors opens up, the voltage divider will still prevent the total voltage potential of the node being from being placed on the test point.

It's also possible to combine two (and sometimes more) analog signals into a common test point signal using a resistor summing network. This will sometimes have a small effect on circuit performance, depending upon the ratios of the resistors, but can normally be easily corrected using Thevenin's theorem.

Some additional analog circuit guidelines that make service operations (and factory test operations, for that matter) easier, include the following:

- Use proper connector types.
- Provide matched impedances.
- Provide nearby ground terminals.
- Break feedback loops:
  - Automatic gain control
  - Automatic frequency control
  - Phase locked loops.

Using proper connector types reduces the number of adapters and connections required for testing. Impedance matching is important in analog circuits and is done differently than normal. Usually an impedance matching network is designed for maximum power transfer. For test purposes, it is designed for minimum interaction with the circuit under test. Finally, just as with digital circuitry, we want to break analog feedback loops to help in definitively isolating faults.

BIT DIAGNOSTIC CONSIDERATIONS

When BIT is to be used successfully to reduce the amount of external test equipment required for service operations, it is wise to select circuits during the product development phase that have on-chip "maintenance mode" control lines that provide loop-back of I/O signals for test purposes. The same goes for things like relays—types exist that send a separate signal out to indicate positive activation. If on-chip loop-back is not available, extra circuitry, under control of the BIT controller, will have to be added, as illustrated in Figure 9-7.

Implementing digital BIT is usually a relatively straightforward task. The presence of analog circuits, however, can complicate that task. Shown in Figure 9-8 are two possible methods for converting analog signals to digital data. One is a simple comparator and the other actual digital-to-analog and analog-to-digital converters. The method of choice depends on the amount of space available, the parts costs trade-offs, and the BIT accuracy requirements. In any case, provision should be made for inserting errors into BIT circuitry (such as via the V_F line, as shown in the example, or by having a known signal on the board that can be switched in) so that the BIT circuitry itself can be tested.

There are several other methods for converting analog signals to digital signals for BIT purposes. Voltage-to-frequency and frequency-to-voltage converters can be used. They (and the comparators and converters just discussed) can be fed via analog multiplexers or FET switches to multiply their usefulness and limit their number. Sensors and transducers can be used to convert electromechanical and electro-optical signals. There should also be provision for switching in a known parameter in order to verify the functionality of the BIT circuitry itself.

With the advances to date in chip technologies, it is possible to build the

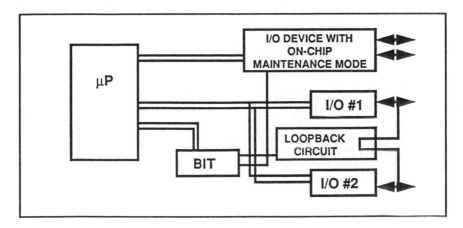

FIGURE 9-7   I/O loopback

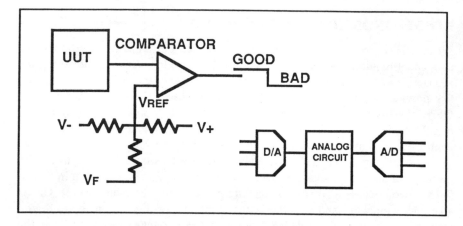

FIGURE 9-8   Converting analog to digital for built-in test

equivalent of an entire IBM-type PC/AT with only a handful of chips. In large systems, such a chip set can be used to provide an embedded BIT controller, as illustrated in Figure 9-9. Or a PC-type plug-in board can be used instead.

Figure 9-9 illustrates how such a chip set would be used. Notice that it is connected via a second set of lines to the system CPU and the other subsystems. In a multi-board system, adding a handful of chips to make the product eminently serviceable can be done without much overhead in terms of parts count or system physical space. The cost will be recouped many times over. Customer satisfaction and uptime will also be increased.

Can you tack this type of capability in after the main functional product design

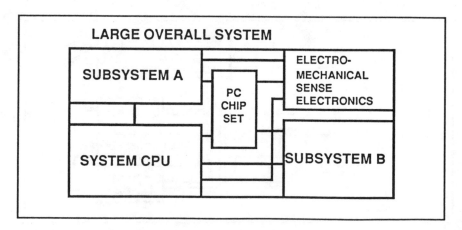

FIGURE 9-9   Embedded PC-based built-in test controllers

has been pretty well completed? Not very often. But if we concurrently engineer it, we might even make the BIT processor do "double duty" and add some extra features to the product as well. Thus, it is important that BIT and service experts be part of the design team right from the start of a project.

REMOTE DIAGNOSTICS

Combining ideas like testability, built-in test, on-board service memory, and embedded testability bus cards and BIT controllers with the communications capabilities of a modem lead to the scenario depicted in Figure 9-10.

If we use this approach, we no longer have to send people out every time to repair systems. For you service engineers who may be worrying about your jobs by now, don't worry—be happy! Change from a service engineer to a serviceability engineer and get involved in the concurrent design aspects of new products.

SUMMARY

There is a lot of commonality between the testability guidelines and the serviceability guidelines. Some of the relationships between the manufacturing and service test operations include the following:

- Basic partitioning, control, and visibility help both.
- Built-in test can reduce manufacturing test engineering efforts.
- Commonality helps with correlation and repair of field returns.
- Transportable tests reduce software development efforts.
- Concurrent development speeds time to market.

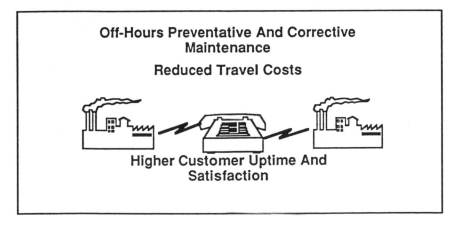

FIGURE 9-10   Remote diagnostics

If the service features of a product are concurrently engineered, a lot of the features put in to help manufacturing will automatically help service. And a lot of the "stuff" required to enhance serviceability mentioned in this text will already be there based on enhanced manufacturability and manufacturing level testability.

Design for serviceability is no more difficult than any of the "design for..." items mentioned throughout this text. The key to success—shorter time to market, higher product quality, lower overall business costs, and increased customer satisfaction—is the *concurrent implementation* of all of the design features that will make sure that a product is successful from all business aspects—performance, market share, price, and profits.

REVIEW

1. Serviceability features are (a) *added to a product design* or (b) *incorporated into a product design.*
2. There are (a) *one*, (b) *two*, (c) *three*, (d) *four*, (e) *five*, or (f) *six* elements that design for serviceability encompasses.
3. Successful built-in test implementation (a) *does* or (b) *does not* require loop-back circuitry.
4. Partitioning for serviceability should be accomplished with (a) *electrical means*, (b) *mechanical means*, or (c) *both electrical and mechanical means, depending upon the product and the service philosophy.*
5. If concurrent engineering processes are employed, there (a) *is* or (b) *is not* a lot of commonality between the BIT functions required by the factory, the field, and the customer.

Answers: 1: b, 2: f, 3: a, 4: c, 5: a

NOTES:

10

Tying It All Together

In the previous chapters, we've discussed why we should do concurrent engineering, the time to market and economic advantages of doing so, the elements of concurrent engineering, and the specific technical details that should be considered in each area (i.e., performance, manufacturability, testability, and serviceability).

In this last chapter, we're going to look at *how you can make concurrent engineering happen in your organization.* We're going to look at reality as it is, at reality as it should be, and how you can be an agent in changing today's condition to a condition that will allow you (and the organization you choose to be associated with) to take advantage of the concurrent engineering discipline.

Concurrent engineering is not, by itself, a new concept. Many people in many companies are practicing some of it in some fashion. They have learned by experience that time and money spent up front can reap large rewards later on. With recent publicity, however, it is gaining impetus rapidly in the form presented in this text and in the LSTI-produced Concurrent Engineering Management™ Session, the regular Concurrent Engineering Seminar™ and the Concurrent Engineering Technical Session™.

As stated in Chapter 2, our experience has been that many product design decisions in organizations that practice serial engineering are made based on opinions, not facts, and that many of those decisions are also made "in a vacuum." We presented the facts of concurrent engineering in Chapter 1. All of the examples are from real companies and have been well documented in the trade and business press. **FACT**: Concurrent engineering can get you to market sooner. **FACT**: Concurrent engineering can make you more cost and price competitive. **FACT**: With the complexity of today's product designs and the fickleness of today's markets, you cannot afford to base design for performance, manufacturability, testability, or serviceability decisions on opinions.

Complexity for complexity's sake is counterproductive. How many of the features of most of your sophisticated electronic products do you (or your customers) *actually use* on a regular basis? Sometimes product simplification can make a product more marketable! That's why you need accurate input from all of the business elements when making design decisions.

People in the manufacturing, test, quality, and service functions often have large amounts of data (i.e., facts) regarding the overall times and costs associated with bringing certain products to market (and their on-going production and warranty/service costs). These facts should be taken into account when new products are designed. We can learn from what we've done right (or wrong!) before.

A reminder to those in manufacturing, test, quality, and service functions: The facts you bring forward must be (a) timely, (b) accurate, and (c) presented in the proper manner. The data you hold gives you power. It must be used wisely—*for improvement, not punishment* of other parts of the organization (or, worse yet, specific people in other groups).

That means saying: "If we do this the same way again, the following will result. If we do it with the concurrent engineering discipline in mind, we have the chance to...," rather than saying "Your last design was a disaster for (manufacturing, test, or service). Look at these numbers...."

Shorter product life cycles and increased pressure for shorter time to market make it imperative that we renounce the "redo-until-right" philosophy and replace it with the *"right-the-first-time" philosophy.* What does "right" mean? The proper set of design trade-offs for the overall success of the product (and the business), given the specific customer requirements, business capabilities, and competitive environment.

A silicon iteration for an ASIC may cost weeks (or even months) in time to market. An iteration can be caused by a lack of communication between the ASIC designer and the system designer. It can be caused by neglecting to fully simulate the operation of the part in the overall product. It can be caused by having to redesign the part to include boundary scan so that manufacturing can test the product that contains the part. It can be caused by inadequate input from product marketing. There are lots of causes (and even more excuses). There is only one prevention—concurrent engineering. And even though its practice won't prevent all of the problems all of the time, you have a much better chance to improve your "hit ratio" when you use it properly.

We showed the impact of design decisions on product life cycle costs early in this text. The earliest decisions usually have the largest impact. As more things get cast in silicon (or in copper on epoxy-glass boards or in the final product package), we have less flexibility in changing things to improve the items that make up the elements of concurrent engineering. Product goals, strategies, and tactics must be planned out as early in the product development cycle as possible—preferably right at the beginning when the product is specified. Those of you in functions that are currently "downstream" from design engineering must take it upon yourselves to get involved in the product design process if you are going to be a source of solutions (rather than an after-the-fact source of problems and complaints).

If the customer requirements dictate a design approach outside the scope of current company capabilities, everyone needs to know about it so that plans can be developed to cope with it. If the design approach can be modified to fit into current company capabilities, so much the better. If neither of those happen, time will be lost, money will be wasted, and market share will be eroded.

It is important to maintain focus on our business purpose so that our activities can best contribute to it. The design decisions that we make and the inputs to design trade-offs that we make should all be focused on achieving the best leverage in reducing time to market, lowering product costs, and increasing product quality.

A product with too few (or too many) features may turn off a customer. A product with too much built-in test capability for its requirements may cost too much. A product with too little may not meet customer specifications. A product that is hard to build creates manufacturing defects, worker stress, and excess costs. A manufacturing process that is obsolete inhibits design innovations that could beat the competition.

The key is *balance* between all of the organizational functions during the concurrent engineering process to achieve the goals that are graphically illustrated in Figure 10-1. Shorter time to market at the expense of cost and quality represents a very short-term mentality and is a very poor bargain in the long run.

As illustrated in Figure 10-2, different elements of organizations have different "flavors" and different requirements and desires. The chart illustrates some of the differences in typical mentality among typical segments of an overall business.

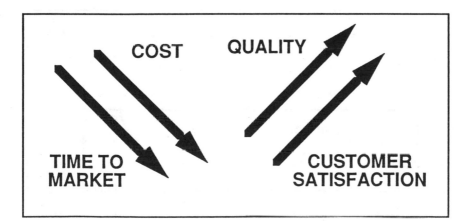

FIGURE 10-1   Concurrent engineering goals

True concurrent engineering requires management to balance the differing needs of different portions of the organization. The idea is to design the project in parallel with the process(es) so that the things that get built can be built in a manner structured enough for the process—to make things flexible enough so that relationships, strategies, and tactics can be tailored to the customer while still providing for structure in the engineering and manufacturing functions.

Marketing and sales function's "musts, needs, and wants" must be balanced with engineering and manufacturing functions that lend themselves to checklists and design reviews (i.e., structured versus flexible approaches). This is not always easy, which is why there are senior management positions filled with those who must manage that balance. The only real common language between all of the diverse elements of an organization is the language of economics. This is why one of the concurrent engineering commandments is to base design decisions on quantitative economic trade-offs.

QUANTITATIVE DESIGN

Chapters 7 through 9 provide many guidelines on improving the manufacturability, testability, and serviceability of a product in this seminar. Which ones do you use? To answer that question, you must estimate the costs of the various alternatives. The earlier in the design phase you do the estimating, the better. Remember also that we are not talking just design cost—we are talking overall product cost and overall business costs. Estimate the cost of each alternative on each phase of the product's life—development, production, and maintenance.

Don't forget the impact of time to market on profits. You will recall that being

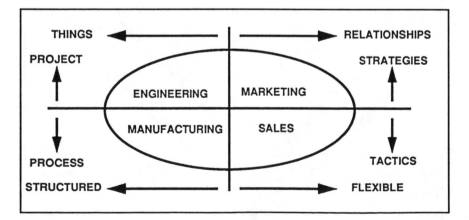

FIGURE 10-2   Organizational orientations

six months late has a significant impact over the life of the product. We want to avoid any glitches or redesigns to get to market as quickly as possible with the highest quality product that has the lowest overall cost.

We said "design quantitatively" and "estimate costs" early in the design phase. How do you do those tasks? How do you quantify all of the cost data needs that were identified in Chapter 2? We suggest the "flow chart and formula" approach. The flowchart for design is shown in Figure 10-3.

The formula that goes with the flow chart of Figure 10-3 is:

$$\text{Total Time} = \text{Design Time} + \text{Review Time} + (F_1)(\text{Redesign Time} + \text{Redesign Verification} + \text{Review Time})$$

You can actually calculate the cost of any redesign and compare it to the cost of taking a little more time during the initial design to prevent redesign activities. Knowing the cost of a redesign lets you know exactly how much extra design time you should spend, making sure that the concurrent engineering process is complete. But don't look at design cost in a vacuum. Look at in the context of the overall business cost of the product.

You can use the flow chart and formula approach to estimate virtually any cost. The flow chart shown in Figure 10-4 is for functional testing of boards or completed products. The formula associated with it is:

$$\text{Total Cost} = \text{\$/Hr}\,[HT + TT + (F_1)(DT + RT + HT + TT) + (F_1)(F_2)(DT + RT + HT + TT) + \ldots]$$

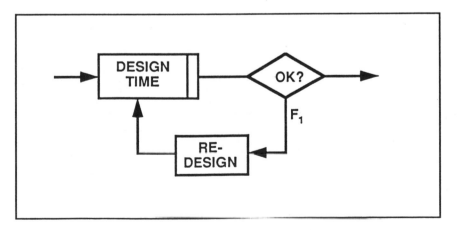

FIGURE 10-3   Flowchart for design

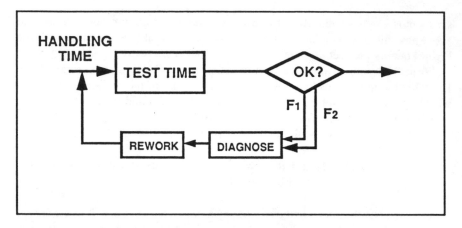

FIGURE 10-4   Flowchart for test

With the labor rate known for each person performing each operation, it is possible to calculate the costs. You can then play "what if" with each of the major parameters and compare alternative strategies. With the total quantity of items to be produced, you can look at the life cycle costs for the various alternatives. Estimating the costs of each alternative can be done and *you can do it!*

Once you have developed the flowcharts and formulas for each of the specific operations that must be performed, add them all together, as shown in Figure 10-5, to get the total cost. We suggest building a spread sheet on your computer so that you can play "what if" with speed and ease and so that you can compare the various options available to you.

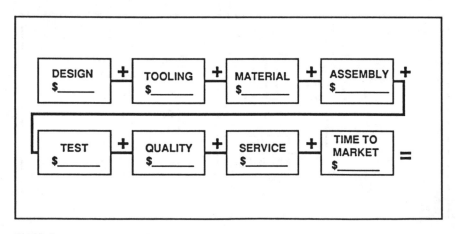

FIGURE 10-5   The life cycle flowchart

The totals that you get using the flow chart and formula approach are the **recurring** costs associated with each operation. Remember to add the nonrecurring costs (e.g., capital equipment, manufacturing, and test software and fixturing development) to them to get the real totals. Don't forget the extra (or lost) profit potential due to time to market and market share. They may feel intangible, but they can be quantified.

But what if you don't personally know how to estimate the costs for areas outside your specific engineering discipline? That's why you have a team! *Each team member is responsible for estimating the costs associated with his or her area of expertise.*

CONCURRENT ENGINEERING AS A PROCESS

Keep in mind that concurrent engineering is an ongoing process, not an instant fix program. It requires top-down specifications and bottom-up design activities. Creating top-down specifications will get easier the more often people do it and see the benefits of doing it. Designers will implement the concurrent engineering technical guidelines during bottom-up design more often when they realize that it doesn't have to take a lot of time or cost a lot of money and that doing so will shorten time to market and simplify their own design verification tasks.

Management must be both committed and involved. And you must carry the concurrent engineering message to the managers in your organization. You can do it. You have that power. Your colleagues need education. Let them know that in a nice way, educate them informally yourself, or arrange to have LSTI bring its Concurrent Engineering Program™ in-house, tailored to your exact needs.

Remember to communicate regularly with both your internal and external customers, for they pay your salary. Lastly, recognize that implementing the concurrent engineering process in your organization takes commitment. Without it, there will be no results.

OVERCOMING INERTIA

That about covers the what, why, and how of concurrent engineering. You know the kind of positive impact that concurrent engineering can have on your company (and even possibly on your career). You know its importance in terms of building, maintaining, or regaining a competitive posture in world electronics markets. You have many pages of information on how to make your new designs easier to verify, build, test, and service. You know how to estimate the overall life cycle costs of a product and how to make the right concurrent engineering trade-offs. Also, we've addressed how to overcome most of the traditional problems. What's left but to go to work and implement what you've learned?

Have either of the following thoughts run through your mind as you've read this book: "This all makes sense, *but* (a) *I can't* because....or (b) *they* won't let me...? They usually do. I mean, this is all good stuff, but I live in the real world. I can't make this happen. 'They' won't let me."

We are now going to take a radical departure from the competitive, economic, management, environmental, and technical topics that have made up the bulk of this book. We are going to discuss people, authority, and power. We are going to discuss permission and forgiveness. We are going to discuss risk taking. We are going to discuss "comfort zones." When we are finished, you will have the knowledge that you need to make concurrent engineering happen in your organization.

Because, when you say "I can't because...," you will be right 100 percent of the time. Why? Because you won't try! "I can't because..." is the biggest "cop-out" in use today. I can't because of budget constraints. I can't because of foreign competition. I can't because my boss won't let me. I can't because the designers don't listen to me. I can't because...

Have you ever heard a "world class" anyone, be it a business leader or an athlete, say that they won because they didn't think that they could? Of course not. World class people have an "I can" attitude. More importantly, they have an *"I will"* attitude. If you think you can't accomplish something, you won't. If you concentrate on negatives, that's what you'll engender. If, on the other hand, you concentrate on positives and work toward their achievement, you cannot fail.

"They won't let me..." is the second biggest "cop-out." There are no "theys." There are only "hims" and "hers" to be persuaded to change their behavior. People just like you and me. We *are* they!

Everyone fears change. It's normal to fear it. Sometimes its better to resist change and sometimes it is not. When it comes to improving your competitiveness through the concurrent engineering process, it is better to change than to resist. The handwriting is on the wall for those who do not adapt. Behavior can be changed in the short term via fear and pain. For long-term changes, pleasure and benefits work much better. You should by now be familiar with the benefits that can be obtained with concurrent engineering. They are significant. They are real. Not changing will lead to fear and pain!

Ah, still makes sense, doesn't it? But what can you, just a cog in the wheel, really do? You can, in fact, have a tremendous impact. None of us are nearly as helpless as we may sometimes feel. Not if we really understand the difference between authority and power.

AUTHORITY AND POWER

Authority, for those of you who have not had it explained to you recently, is something that someone, usually a boss, gives to you or bestows upon you. You may be given authority by virtue of the title of the office you hold or by virtue of the position itself. You may be given, for example, the authority to hire people, to terminate people, or to spend company funds (usually up to a certain "authorized" limit). Since authority is given to you, usually by someone higher up in the organization (who may be perceived as more powerful, but who actually has only had more authority granted to him or her), it can easily be taken away.

You probably don't have the "authority" to cause concurrent engineering to be implemented in your organization. But you do have the *power* to do it. Power is a different animal altogether. Where does power come from? Who gives it to you? It turns out that no external person gives you power. It comes from within. You give it to yourself. You alone. No "theys." No hims nor hers. Just you.

You have within you all of the power you need to make concurrent engineering a reality in your organization. All you need to do is give yourself permission to use it. Since power is not bestowed upon you by someone else, it can never be taken away. Since power is so absolute, it carries both responsibility and risk. This means that you need to assess each situation carefully before exercising your power. Look at the risk-reward ratio. Examine the possible consequences of your contemplated actions. Make sure that there will be mutual benefit. But never doubt that you have the power to effect change.

In the final analysis, the decisions you make are yours and yours alone. But you do have the power to accomplish anything that you want badly enough to accomplish.

VISUAL IMAGE FORMATION

We're now going to talk about a tool that you can use in your everyday life to help you effect change. It's not limited to organizational or procedural change—it works for anything. Someone once said: "Whatever you vividly imagine, ardently desire, sincerely believe in, and enthusiastically act upon must inevitably come to pass." That statement is true.

Visual image formation is not some strange California-based cult! It is a proven success tool that you can use to achieve your goals, whatever they may be. Athletes use it. Business leaders use it. Books have been written about it (The Power of Positive Thinking, The Magic of Believing, etc.).

You want a new house? You want a new car? You want concurrent engineering? In the words of Star Trek's Captain Picard: "Make it so."

Visual image formation works by putting the power of your subconscious brain to work for you. Most scientists and psychologists agree that we use only about 10 percent of the power of our minds—the conscious part—most of the time and that the other 90 percent of it, the subconscious, is vastly under-utilized.

The conscious portion of our brain works well with complex words, figures, and concepts. We use it for reasoning and logic. The subconscious, on the other hand, does not deal well with complex words, figures, and concepts. It responds much better to simple concepts and visual images. So in order to communicate with it, to put it to work for us, we must use either very simple (one sentence) "I will...because..." type statements or, preferably, visual images (i.e., mentally created pictures, in full color and three-dimensional).

We all have certain mental images of ourselves, our jobs, our relationships with others, and our environments as they currently exist. We also have certain mental images of how we would like reality to be different from the way it is now. We sense the way reality is now with our physical senses (touch, sight, hearing, etc.).

If we are currently living life exactly the way we want to, the reality that we visualize (i.e., our mental image of reality) exactly matches reality as we presently physically sense it. If we are not, however, reality as we visualize it and reality as we sense it are not equal—they do not match. One of the jobs that the subconscious mind is best suited for is resolving differences between reality as we visualize it and reality as we sense it. It will either make our mental images match today's reality or, more productively, make us do the things that are necessary to make reality as it is today begin to match our mental image.

When reality as we sense it and reality as we visualize it do not match, disharmony is created. The subconscious mind does not like disharmony. It doesn't feel good. But the subconscious mind is extremely creative. It can be a very powerful force for resolving discrepancies or disharmony, and it will do so in one of two ways:

- It will alter the mental image to resolve the disharmony, or
- It will give you ideas (or even direct you) to do the things that are necessary to make present reality match the visual image that you have created.

Think about it. How many times have you "slept on" a problem, only to awaken with a solution? That was your subconscious mind resolving it for you. Whenever disharmony (sometimes also called tension) exists, your subconscious mind will go to work to eliminate it.

There are two types of disharmony—distracting, or negative, disharmony and positive, or stimulating, disharmony. An example of distracting disharmony is the noise of a printer in an adjacent area while you are trying to concentrate on something requiring your full attention. What happens to the noise as you really begin to focus your conscious mind on the task at hand? It disappears, doesn't it?

The noise was creating disharmony. So your subconscious mind "turned off" your ability to hear the noise until you made a conscious decision to stop concentrating on the task at hand and reemerge into the fullness of your environment, printer noise and all.

Positive disharmony works the same way, except that instead of allowing the subconscious to "turn off" a physical sense in order to eliminate the disharmony, we deliberately create disharmony and let the subconscious mind go to work and make us do the positive things to resolve the disharmony. How do you create positive disharmony? It's really quite simple. You make a clear mental picture (i.e., form a solid visual image) of reality *the way it should be* for you. You keep that picture constantly in your conscious mind, couple it with a good feeling from your past, and transmit it repeatedly to your subconscious mind. That's all you have to do. Your subconscious mind will go to work for you and give you the drive, the ambition, the ideas, and the solutions that will make reality as it is match reality as it should be (i.e., as it is in your visual image). It will thus resolve the disharmony for your benefit.

It's important to couple the visual image with an emotion—a good feeling from an event or activity that you particularly enjoy(ed). That way, the creative subconscious mind will know that, when it gives you what you need to resolve the disharmony, it will *feel good*. Formulate the picture as if the successful result *has already happened*. There must be no doubt in your mind as to the validity of either the image or the feeling attached to it. Send them both to the subconscious as if they were a foregone conclusion.

Does all of this sound too simplistic and "off the wall"? You might want to think again. Here's an example of how it has personally worked for one member of LSTI's staff.

One of our people had an abstract for a paper accepted for an ATE conference in England. He then had to write the paper, put the slides together, and give the talk for a 300-person audience from multiple countries. Having been to the conference before, he knew about the "Best Paper Award" and decided that he was going to win it.

He had seen another person get the award two years earlier. So he made a mental picture of that ceremony with himself in it. Receiving the trophy following a captivating talk. The picture was made before the paper was written and before the slides were made. Every time he worked on the paper or the slides, he remade that picture and refelt how good it was going to feel *when* (not if!) he won. He also remade that picture, coupled with the feeling, every time he rehearsed the talk and immediately prior to giving it. He won the award.

Visual image formation works. Believe it and put it to work for you. There is nobody stopping you but you! Imagine not what it would feel like if you can successfully get concurrent engineering implemented in your organization. Imagine instead what *it will feel like when you accomplish it.* For if you truly believe that you can make the reality match the picture, you will. Your creative subconscious mind will go to work to make you do the things that are necessary to make reality match your mental image.

What did our author/speaker do that was so special? On a conscious level, he doesn't know. He just "knew" he was going to win and that (self-created) knowledge flavored the development of the paper, the slides, and the presentation. And he won.

Make a picture of how concurrent design should be done in your organization. *Make it real and make it feel.* Then you'll be able to make it happen.

PERMISSION VERSUS FORGIVENESS

There's an old saying, the origin of which we're not sure of, that goes like this: "It is much easier to gain forgiveness than it is to get permission." And that old saying contains more than a small element of truth. If you reflect upon the people you consider to be successful "movers and shakers," isn't one of their character traits the ability to do things and then explain what they did and why they did it after the fact? Sure it is.

If your "unauthorized" course of action doesn't succeed, you can always loudly proclaim your responsibility, state that you are definitely to blame, tell everyone that you've certainly learned your lesson this time, and ask for forgiveness. Ninety nine times out of one hundred, unless you have done something that intentionally damages the organization, done someone grievous harm, or done something legally or morally wrong, you'll be forgiven.

Are you going to be fired for saving the company a lot of money, getting its products to market sooner, and increasing its market share? Highly unlikely. If you do get censured for improving things, it's time to look at whether or not you want to continue to work there!

The drawing in Figure 10-6 illustrates the integration of new knowledge into your "comfort zone." Notice how the circle on the right side of the figure has grown larger, more powerful. For knowledge is power.

Is there anyone in your organization who has not read this book, regardless of their level of authority, who can currently claim to know more about the concurrent engineering discipline than you? Do you now need their "permission" to implement the appropriate concurrent engineering guidelines in your next design? Do you need their "permission" to let everyone involved with new designs know how to cut time to market, reduce product costs, and increase market share?

Of course you don't. You have the power to do all of the above without anyone's permission. Especially since it is for the good (or even survival) of the organization that you work for.

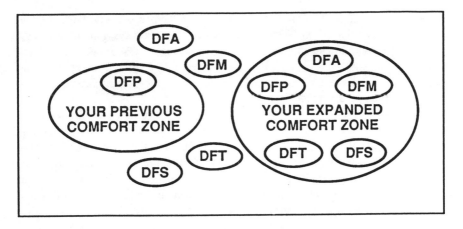

FIGURE 10-6   Comfort zones

Not everything can be accomplished instantly. But everything can be accomplished eventually. And timing is very important. We're going to look at one last example of how (and when) to use the power that you have to effect the necessary changes in your current design process.

Let's suppose that you're in a large, important meeting on corporate strategies, costs, and time to market issues. Let's further suppose that an appropriate moment arrives (i.e., the division GM might have just said something about "cost of goods sold being too high") and that you have something to contribute. Something such as, "You know, if we were to implement the concurrent engineering discipline on the next project, we could cut the cost of goods sold by 10 to 30 percent."

Recognize that moment as a significant opportunity and seize it. Do not be shy. The meek may someday inherit the earth, but what's left of it will be up to the bold. *Stand up!* The moment you stand up, you have taken control of the meeting. Remember, *you have the power to stand up.* Your power exceeds the authority of whoever is currently leading the meeting.

Walk to the front of the room, saying something like, "Excuse me, but you've just triggered a thought that I think will be beneficial to this discussion, and I'd like to explain it right now while it's both fresh in my mind and germane to the matter at hand." Invade the meeting leader's "comfort zone." Put your hand out for the chalk (or the felt pen, etc.). You will get it. Say your piece briefly—remember that this is a management meeting. Say "thank you" and sit down.

Some of your associates may think that you are crazy to take such a risk. So what? If you are right (and you are!), a firm decision to improve things will be made on the spot and you'll have been instrumental in making it happen. Also remember that your subconscious will do whatever is necessary to resolve disharmony. This includes automatically creating "filters" or "blind spots" as time goes by. We tend to go from "I want that fixed *now!*" to "When will that be fixed?" to "Will that ever be fixed?" to "What needs fixing? We've always done it that way."

The filters and blind spots must be consciously removed. We need to step back now and then and make sure that we have not let old habits creep back in. Continuous improvement and eternal vigilance are the watch-phrases for the concurrent engineering discipline.

Reality is what is. It is the way things are. It may be right, it may be wrong. It may be comfortable, it may not be comfortable. But it is what it is. It is critical that you acknowledge the realities of the people, the products, the organization, and the politics around you if you are to create the right objectives, strategies, and tactics for achieving your goals. Realities such as which people are in the real power positions, what would make them want to help you, and what you need to do to persuade them to your point of view. But it is also critical that you recognize reality as *something that you can change.*

You now have both the technical and the motivational information you need to get the concurrent engineering process implemented in your new designs. You can, and indeed must, make it happen. You have that power. If our organization can help in any way, please don't hesitate to contact us. We'll do whatever we can to help you with either in-company seminars or specialized consulting services.

SUMMARY

The practice of concurrent engineering allows an organization to make the fullest use of its best and brightest people to bring higher quality, lower cost products that increase customer satisfaction to market sooner. It's focus on teamwork and team accountability breaks down traditional organizational barriers and helps management achieve the balance that has long been missing in the serial design environment. By applying all of the right expertise at the right time, organizations can continuously improve the quality of their products and indeed of their businesses as a whole.

Creating multifunctional and multidisciplinary teams and empowering them with the tools and information they need for quantitative design is part of the total quality management effort that has been so successful in many factories. It is time to use those same principles in the area where they can have even bigger impact— during product design.

In addition to the obvious benefits of concurrent engineering that were pointed out early in this text, some subtle benefits also come into play. These subtle (or sometimes not so subtle, depending on your point of view) benefits include:

- Lower capital equipment cost
- Greater use of automation

- Less chance of redesign
- Fewer parts to buy from fewer vendors
- Better factory availability
- Improved design quality
- Improved organizational motivation and morale

These additional benefits position a company to respond ever more quickly to changes in the marketplace and to take rapid advantage of new technologies—both product and process.

REVIEW

1. The concurrent engineering philosophy is (a) *redo until right* or (b) *right the first time*.
2. Concurrent engineering is (a) *a quick fix program* or (b) *a long-term process*.
3. Concurrent engineering is (a) *capital intensive* or (b) *communication intensive*.
4. It is usually easier to get (a) *forgiveness* or (b) *permission*.
5. The responsibility for getting concurrent engineering implemented in your organization lies with (a) *you*, (b) *your boss*, or (c) *someone else*.

Answers: 1: b, 2: b, 3: b, 4: a, 5: a, then b, then c!

NOTES:

Appendix

The following checklists are provided to help in the concurrent engineering process:

- Concurrent Engineering Management Checklist
- Concurrent Engineering Team Checklist
- Concurrent Engineering Technical Checklists

Permission is granted to the reader to make unlimited copies of these checklists for internal use only. All other rights under U.S. and foreign copyright laws reserved.

Note to the reader: As part of the Concurrent Engineering Seminar™ series, a mathematical spreadsheet called the Concurrent Engineering Spreadsheet™ has been developed. If you would like to receive information on either the seminar series or on the spreadsheet, contact Logical Solutions Technology Incorporated, 96 Shereen Place, Suite 101, Campbell, CA 95008 USA, Telephone 408-374-3650, FAX 408-374-3657.

Concurrent Engineering Management Checklist

Needed		Completed
_____	Market Research	_____
_____	Business/Product Plan	_____
_____	Concurrent Engineering Mission Statement	_____
_____	Concurrent Engineering Team Assignments	_____
_____	Concurrent Engineering Team Education	_____
_____	Product Specifications/Requirements	_____
_____	Physical Facilities In Place	_____
_____	Communication Methods In Place	_____
_____	Design Automation Resources In Place	_____
_____	Non-Recurring Cost Data Available	_____
_____	Recurring Materials Cost Data Available	_____
_____	Recurring Assembly Cost Data Available	_____
_____	Recurring Test Cost Data Available	_____
_____	Expected Production Yield Data Available	_____
_____	Forecast Production Volumes Available	_____
_____	Goals Set And Agreed Upon By Team	_____
_____	Schedules Set And Agreed Upon By Team	_____
_____	Performance Criteria In Place And Agreed	_____
_____	Supplier/Customer Non-Disclosures Signed	_____
_____	Project Progress Review Schedule Set	_____
_____	Project Design Review Schedule Set	_____
_____	Regulatory Requirements Identified	_____
_____	Manufacturing Strategy/Plan In Place	_____
_____	Test Strategy/Plan In Place	_____
_____	Service Strategy/Plan In Place	_____
_____	Documentation/Customer Training Plan In Place	_____
_____	Product Development Budget Approved	_____
_____	Project Monitoring Methods In Place	_____
_____	Other _____	_____
_____	Other _____	_____
_____	Other _____	_____
_____	Other _____	_____

Concurrent Engineering Team Checklist

Needed   **Assigned**

Needed		Assigned
_____	Project/Program Manager	_____
_____	Marketing Team Member(s)	_____
_____	System Design Team Member(s)	_____
_____	Design Automation Team Member(s)	_____
_____	Packaging Design Team Member(s)	_____
_____	Industrial Design Team Member(s)	_____
_____	Components Group Team Member(s)	_____
_____	Device Design Team Member(s)	_____
_____	PCBA Electrical Circuit Design Team Member(s)	_____
_____	PCBA Layout Team Member(s)	_____
_____	System Software Team Member(s)	_____
_____	Accounting/Finance Team Member(s)	_____
_____	Documentation Team Member(s)	_____
_____	Manufacturing Engineering Team Member(s)	_____
_____	Test Engineering /BIT/BIST Team Member(s)	_____
_____	Maintainability/Service Engineering Team Member(s)	_____
_____	Thermal Analysis Team Member(s)	_____
_____	Environmental Analysis Team Member(s)	_____
_____	Regulatory Compliance Team Member(s)	_____
_____	Human Factors/Safety Team Member(s)	_____
_____	Reliability Team Member(s)	_____
_____	Mechanical Design Team Member(s)	_____
_____	Quality Assurance Team Member(s)	_____
_____	Purchasing Team Member(s)	_____
_____	Materials Team Member(s)	_____
_____	Customer Team Member(s)	_____
_____	Supplier Team Member(s)	_____
_____	Management Team Member(s)	_____
_____	Other _____ Team Member(s)	_____
_____	Other _____ Team Member(s)	_____
_____	Other _____ Team Member(s)	_____
_____	Other _____ Team Member(s)	_____

Concurrent Engineering Technical Checklist (1)

Reviewed   **Agreed**

Reviewed		Agreed
_____	Total Parts Count	_____
_____	Total Part Type Count	_____
_____	Re-Usable Modules	_____
_____	Standard Components	_____
_____	Standard Packages	_____
_____	Screws, Screw Types	_____
_____	Fasteners	_____
_____	Settings And Adjustments	_____
_____	Sequential Assembly	_____
_____	Self Aligning Parts	_____
_____	Power Supplies And Line Cords	_____
_____	Internal Cables	_____
_____	Physical Modularity	_____
_____	Accessibility	_____
_____	Weight	_____
_____	Connectors	_____
_____	Standard Board Sizes And Grids	_____
_____	Layout Constraints	_____
_____	Tooling Holes	_____
_____	Component Placement/Clearances	_____
_____	Component Orientation/Identification	_____
_____	Restricted Areas	_____
_____	Test Pads	_____
_____	Soldering Considerations	_____
_____	Automated Assembly Considerations	_____
_____	Assembly Equipment Constraints	_____
_____	Other _____	_____
_____	Other _____	_____
_____	Other _____	_____
_____	Other _____	_____
_____	Other _____	_____
_____	Other _____	_____

Concurrent Engineering Technical Checklist (2)

Reviewed   **Agreed**

Reviewed		Agreed
_____	Overall Electrical Partitioning	_____
_____	Overall Electrical Control	_____
_____	Overall Electrical Visibility	_____
_____	Integrated Built-In Test	_____
_____	Manufacturing Test Strategy	_____
_____	Service Test Strategy	_____
_____	Circuit Functional Partitioning	_____
_____	Circuit Type Partitioning	_____
_____	Logic Level Partitioning	_____
_____	Feedback Loops	_____
_____	Timing Diagrams	_____
_____	Remote Diagnostics	_____
_____	Dual Port Access	_____
_____	Initialization	_____
_____	Clocks And Oscillators	_____
_____	One Shots	_____
_____	Buffers	_____
_____	Wired-OR/Wired-ANDs	_____
_____	Counters And Shift Registers	_____
_____	Programmable Logic	_____
_____	LSI/VLSI Logic Control	_____
_____	LSI/VLSI Board Visibility	_____
_____	On-Board/On-Chip ROM	_____
_____	Asynchronous Circuitry	_____
_____	Bus Condition/Direction Control	_____
_____	Synchronization	_____
_____	Software Initialization	_____
_____	CISC And RISC Processor Control	_____
_____	Test Access To Internal Nodes And Busses	_____
_____	Other _____	_____
_____	Other _____	_____
_____	Other _____	_____

Concurrent Engineering Technical Checklist (3)

Reviewed   **Agreed**

_____	SMT Board Size Considerations	_____
_____	SMT Board Component Mounting	_____
_____	SMT Board Test Pad Sizes	_____
_____	SMT Board Test Pad Placement	_____
_____	Multiplexed Test Points	_____
_____	Shift Registers For Test Point Access	_____
_____	On-Board Diagnostic Features	_____
_____	Built-In Test Approaches	_____
_____	ASIC Internal Scan Paths	_____
_____	ASIC BIST/BILBO Features	_____
_____	ASIC Boundary Scan Features	_____
_____	Board Level Boundary Scan	_____
_____	Scan Chains For Partitioning	_____
_____	Mulitplexed Scan Chains	_____
_____	Standard Testability Bus Interfaces	_____
_____	Testable Functional Circuits	_____
_____	Diagnoseability	_____
_____	Service Accessibility	_____
_____	Component/Subassembly Replaceability	_____
_____	Component/Subassembly Repairability	_____
_____	Component/Subassembly Rebuildability	_____
_____	Product Upgradeability	_____
_____	Subassemblies On Slides	_____
_____	Service Loops On Cables	_____
_____	Accessible Test Points	_____
_____	Provision For Extender Cards	_____
_____	Strain Reliefs On Connectors	_____
_____	Other _____	_____
_____	Other _____	_____
_____	Other _____	_____
_____	Other _____	_____
_____	Other _____	_____

Concurrent Engineering Technical Checklist (4)

Reviewed   **Agreed**

Reviewed		Agreed
_____	Functional Modularity For Service	_____
_____	Enclosure Considerations	_____
_____	Physical/Mechanical Service Partitioning	_____
_____	Analog Signal Interfaces	_____
_____	Analog Metering	_____
_____	Analog Partitioning	_____
_____	Analog Test Points	_____
_____	Proper Connector Types	_____
_____	Impedance Matching Networks	_____
_____	Nearby Ground Terminals	_____
_____	Analog Feedback Loops	_____
_____	Live System Board Replacement	_____
_____	Diagnostic Accuracy Considerations	_____
_____	Test Software Considerations	_____
_____	Fault Coverage Considerations	_____
_____	False Alarm Considerations	_____
_____	Built-In Test/ATE Correlation	_____
_____	Converting Analog To Digital For BIT	_____
_____	I/O Signal Loopback	_____
_____	Embedded PC-Based BIT Controllers	_____
_____	Portable Service Processors	_____
_____	Testability Bus Interfaces For Service	_____
_____	EEPROMs On Cards For Service History	_____
_____	Manufacturing/Service Test Commonality	_____
_____	Required Documentation	_____
_____	Other _____	_____
_____	Other _____	_____
_____	Other _____	_____
_____	Other _____	_____
_____	Other _____	_____
_____	Other _____	_____
_____	Other _____	_____

Bibliography

1. Bateson, J. *In-Circuit Testing*. Van Nostrand Reinhold, 1985.
2. Fujiwara, H. *Logic Testing and Design for Testability*. MIT Press, 1985.
3. Goel, P. "Test Costs Analysis and Projections." IEEE Design Automation Conference, 1980.
4. Goering, R. "Boundary Scan Technique Targets Board Level Testability." *Computer Design*, October 1, 1987.
5. IEEE P1149.x Series of Proposed Testability Bus Standards.
6. McCluskey, E. *Logic Design Principles (with emphasis on testable semicustom circuits)*. Prentice Hall, 1986.
7. Parker, K. *Using CAE Tools for ATE Programming*. IEEE Computer Society Press, 1987.
8. Rosenberg, L. "The Evolution of Design Automation Toward VLSI." Proceedings of the 17th Design Automation Conference, IEEE.
9. Tsui, F. *LSI/VLSI Testability Design*. McGraw Hill Book Company, 1987.
10. Turino, J. *Design To Test*. 2nd Edition. Van Nostrand Reinhold, 1990.
11. Anderson, D. *Design for Manufacturability*. CIM Press, 1990.
12. Concurrent Engineering Seminar™, Jon Turino, Logical Solutions Technology Incorporated, 1990.
13. Concurrent Engineering Technical Session™, Logical Solutions Technology Incorporated, 1991.
14. Concurrent Engineering Management™ Session, Logical Solutions Technology Incorporated, 1991.
15. Numerous articles from magazines, including *High Performance Systems, Electronic Design, Electronics Test, EDN, Circuits Manufacturing, Electronic Business, Aerospace Engineering, Defense Electronics, Test & Measurement World, Electronic Packaging & Production, Electronic Products, Evaluation Engineering, EE Times, Business Week, Fortune, Boardroom Reports*, and others whose data has been assimilated and whose identities have merged into my collective consciousness.

Index